Marcela Marçal Alves Pinto Mick
João Luiz Kovaleski
Rui Tadashi Yoshino

THE INFLUENCE OF TECHNOLOGY TRANSFER IN INDUSTRY 4.0

Marcela Marçal Alves Pinto Mick
João Luiz Kovaleski
Rui Tadashi Yoshino

THE INFLUENCE OF TECHNOLOGY TRANSFER IN INDUSTRY 4.0

Proposal of a maturity model

ScienciaScripts

Cover image: www.ingimage.com

This book is a translation from the original published under ISBN 978-613-9-78200-0.

Publisher:
Sciencia Scripts
is a trademark of
Dodo Books Indian Ocean Ltd. and OmniScriptum S.R.L publishing group

120 High Road, East Finchley, London, N2 9ED, United Kingdom
Str. Armeneasca 28/1, office 1, Chisinau MD-2012, Republic of Moldova, Europe
Printed at: see last page
ISBN: 978-620-5-71281-8

SUMMARY

1 INTRODUCTION

Competition among manufacturing companies has become increasingly fierce and in this way, many countries and organizations have proposed new production paradigms to integrate the manufacturing process (KHAN; TUROWSKI, 2016). Industry 4.0 (I4.0) is one of the most popular concepts in advanced manufacturing areas and has been considered as a future direction (WEIGAND; PRAUSE, 2016). Industry 4.0 is the fourth industrial revolution that applies the principles of cyber-physical systems, internet, future-oriented technologies and intelligent systems with human-machine interaction paradigms (FU et al., 2018).

According to Stark (2004), the management of a product from inception to disposal has strategic value for a given company in the networked economy. With intelligent factories and products, changes shall occur in the manner products are manufactured, impacting various market sectors.

In short, Industry 4.0 represents a huge potential in many areas. Its implementation generates impacts throughout the value chain, improving production and engineering processes, enhancing the quality of products and services, optimising the relationship between customers and organisations, bringing new business opportunities and economic benefits, changing education requirements, and transforming the current work environment (PEREIRA; ROMERO, 2017).

However, for the concept of Industry 4.0 to be applied in organisations, there are approaches to processes, operations and activities in the area of Technology Transfer (TT) that must be analysed.

Technology Transfer is considered an active process in which technology is transferred between two distinct entities, whether in the form of knowledge, technology or products (AUTIO; LAMANMAN, 1995; AMESSE; COHENDET, 2001; BOZEMAN, 2000; CORMICAN; O'CONNOR, 2009, BOZEMAN et al., 2015).

According to Zangiacomi et al. (2018), the adoption of approaches for TT plays a relevant role as an enabler in the digital transformation process and is directly linked to the importance of increasing the knowledge base in I4.0 technologies.

The transformation in Industry 4.0 impacts the strategies, organizational structure, operational processes and technological development of a company. Thus, it is necessary that top management supports proactive projects and investments, especially to deal with complex processes and promote market competitiveness (LIN et al., 2020). A Maturity Model (MM) can be used to assess the current state of the organisation, and define its strategies for implementing Industry 4.0. (GÖKALP et al., 2017; AKDIL et al., 2018; SCHUMACHER et al., 2016). Nikkhou, Taghizadeh and Hajiyakhchali (2016) indicated that maturity is a guiding tool to correct and prevent problems and assess organizational attributes, including strengths, weaknesses and opportunities.

There are few works in the literature that relate Technology Transfer with Industry 4.0. Different directions and approaches on these two topics have been found. Gjeldum et al. (2016), Karre et al. (2017) and Posselt et al. (2016) presented on the importance of Learning Factories for Industry 4.0 learning. Other authors have studied the human-machine interface, and how companies relate 4.0 technologies to human resource management (LIBONI et al., 2019; RAUCH et al., 2019; BERDAL et al., 2019; HANNOLA et al., 2018; ANSARI et al., 2020). The university-industry relationship is also presented by different authors (KARRE et al., 2017; CIANFANELLI et al., 2019; KRUGER, STEYN, 2019; BARBOSA et al., 2020; SASTOQUE et al., 2020). Other authors, such as Feng et al. (2017), Mahmood and Mubarik (2020) and Michna and Kmieciak (2020) presented the transfer and/or management of knowledge in Industry 4.0. Studies of the performance of Technology Transfer in the Industry 4.0 Scenario were also found (ITUARTE et al., 2018; DA SILVA et al., 2019; KRUGER, STEYN, 2019; TOURK, MARSH, 2016; FU et al., 2020).

The existing works show that there is a strong impact of TT on I.40. However, none of these works presented what really is the existing influence between these two factors. There are also no works that present models and/or frameworks of this influence, and how this union can be beneficial to organizations.

This study sought to develop meaningful research given the absence of studies of this scope. Thus, the fundamental question that drives the development of this research arises: **How does one determine the influence of Technology Transfer on Industry 4.0 maturity?**

1.1 OBJECTIVES

As a result of the problem presented, the following objectives were outlined.

1.1.1 General Objective

To develop an Industry 4.0 maturity model, where Technology Transfer influences its maturity level in order to facilitate its implementation.

1.1.2 Specific Objectives

The specific objectives of this work, which will make up the general objective, are:

OE1) Elaborate a Conceptual *Framework of* the Influence between Technology Transfer and Industry 4.0;

OE2) Determine the components and dimensions of TT and I4.0;

OE3) Apply the proposed model;

OE4) Analyse which I4.0 and TT dimensions most influence the others and are influenced by them;

OE5) Develop a TT-I4.0 maturity diagnostic for the interviewed companies.

1.2 JUSTIFICATION

The concept of Industry 4.0 is present in companies and academia, however, its features still need to be implemented in organisations. To adopt these new technologies, an evolutionary process is required. Thus, for the organisation to be able to adopt these new technologies, it needs Technology Transfer mechanisms.

The purpose of this research is to propose a model where one can measure maturity levels to evaluate the organization's current Industry 4.0 stage, based on the influence of Technology Transfer with 4.0 concepts and technologies. With this, it is possible to find the measures that might lead them to a higher maturity stage.

This research can be justified as follows: during the systematic literature review, using the *Methodi Ordinatio* created by Pagani, Kovaleski and Resende (2015), it was realized that few articles present the correlation between Technology Transfer and Industry 4.0. Still, it is revealed an absence in the literature of any study that presents a maturity model addressing the influence of these two factors.

The originality of this research is validated by the peculiarities that were not found in other works during the systematic literature review, as demonstrated in Figure 1, proving the uniqueness of the research. By researching separately the keywords *"Technology Transfer", "Industry 4.0", "Maturity Model"* and *"Influence"* in the *Web of Science, Scopus* and *Science Direct* journal bases, a large number of papers are obtained. However, when combined, this number of papers falls circumstantially. Searches were also made in the bases of theses and dissertations in the CAPES Portal to make sure that there were no similar works. The time frame used in this research was 21 years: Jan/2000 to April/2021. The numbers presented in this figure were found before the application of the *Methodi Ordinatio,* after its application the numbers drop considerably. (For this search, the combinations of all variants for the keywords "*Technology Transfer*" and "*Industry 4.0*" were used, Appendix A presents this search in detail).

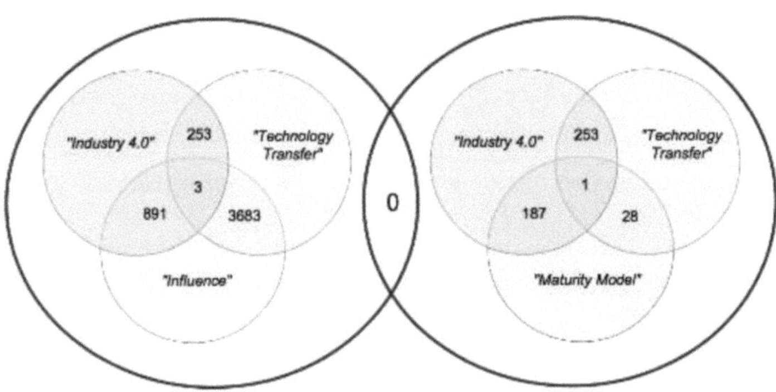

Figure 1 - Survey results
Source: Own Authors (2021)

Although it has been found 01 work that addresses the theme of maturity model of Industry 4.0 with Technology Transfer, and 03 works of influence between Technology Transfer and Industry 4.0, this research remains original for its peculiarities and robustness that are: a) approach of all components of I4.0 and TT in the maturity model; b) elaboration of an unprecedented conceptual framework that identifies the influence between TT and I4.0; c) approach of the influence between I4.0 and TT in a maturity model; d) use of the DEMATEL method to determine the weights of the model dimensions. These factors are not identified in any of the works already published.

The relevance of this work lies in the possibilities of learning for the academic, entrepreneurial and governmental communities generated by the construction and validation of the model. This model will not only result in new knowledge but also assist in future research.

Participating companies benefited from the diagnosis received after the survey, which presents their maturity level of Industry 4.0 and Technology Transfer. Also, according to each company result and scenario, strategies and measures were addressed to take them to a higher maturity stage both in I4.0 and TT, based on the influence of the dimensions.

Different companies can also benefit from this work. This presented model can be applied in any company, and can also be used as a tool to monitor the process of I4.0 and TT implementation.

The approaches presented in this work aim to assist the management of human, financial and technological resources in industries. Thus, they fit into the area of Organisational Engineering, considered one of the ten areas of action of the Production Engineer in the labour market. The work aims to analyse and implement Industry 4.0 concepts in organizations, using Technology Transfer as a facilitator, the main theme of the Research Group to which the researcher is linked.

1.3 STRUCTURE OF THE WORK

This study is structured into nine chapters. The first chapter presents the contextualization of the theme, bringing the objectives and justification of the study.

Chapters 2, 3, 4 and 5 bring the theoretical foundation of this work. Chapter 2 is presenting the main concepts of Industry 4.0, as well as its characterization, process integration and *building blocks*. The main barriers extracted from the studied literatures were also presented. Chapter 3 brings a review about Technology Transfer. Its process, the transfer of knowledge, as well as its mechanisms and the existing barriers are mentioned for a better understanding of the subject. Chapter 4 presents the I4.0 and TT maturity models found in the literature in order to achieve one of the proposed specific objectives. To conclude the theoretical framework, Chapter 5 reports the state of the art, Technology Transfer in Industry 4.0. This chapter is completed with a theoretical model developed by the author, which based on the literature, presents the relationship between TT and I4.0.

The methodology of the work is set out in chapter 6. The chapter is subdivided into sections: Systematic Bibliographic Review and Model Elaboration. The first section describes how the bibliographic portfolio was obtained. In the second section, it is described how the Maturity Model was built in stages.

Chapter 7 details how the application of the I4.0-TT Maturity Model was carried out in the companies, with the intent of validating the same. Chapter 8 presents all the results obtained in the application of this model. Finally, chapter 9 presents the final considerations of this work.

In order to provide an overview of the development of this study, Figure 2 illustrates a flow chart of the steps outlined in this work.

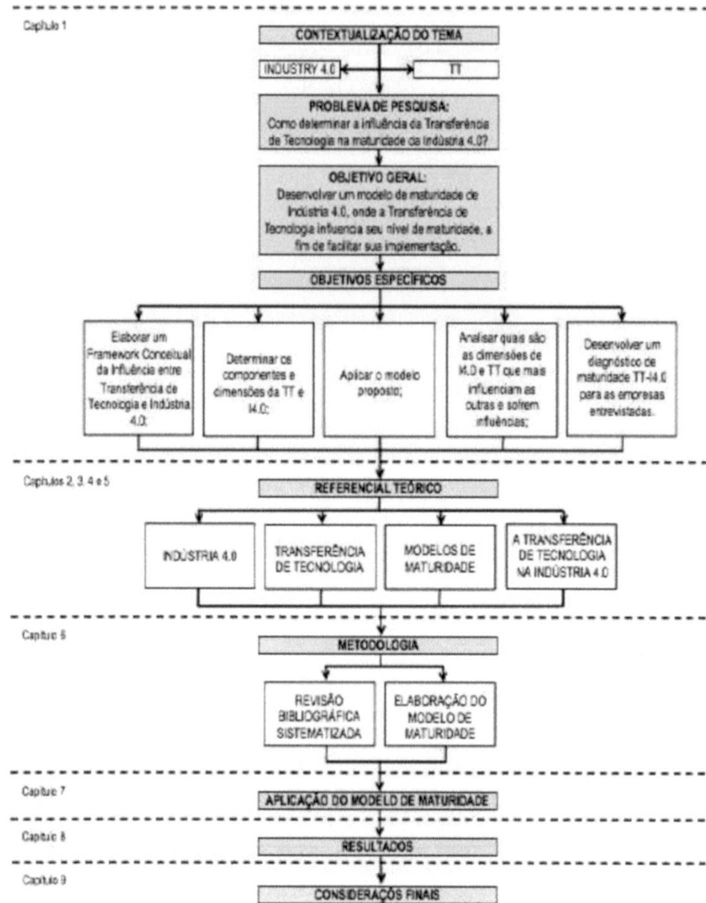

Figure 2 - Flowchart of the work structure
Source: Own Authors (2021)

2 INDUSTRY 4.0

2.1 CONTEXTUALISATION

The First Industrial Revolution occurred in England in the mid-18th century and was enhanced by the invention of the steam engine. During the second half of the 19th century, the Second Industrial Revolution broke out in Europe and the USA. This revolution was characterised by mass production and the replacement of steam by chemical and electrical energy. To meet the growing demand, several technologies in industry and mechanization were developed, such as the assembly line with automatic operations, allowing increased productivity. The invention of the integrated circuit (*microchip*) was the technological breakthrough that triggered the Third Industrial Revolution. The use of electronics and information technology, in order to achieve greater automation in production, are the main characteristics of this revolution that emerged in the last years of the twentieth century in many industrialized countries around the world (ACEMOGLU, 2002; TUNZELMANN, 2003).

According to Schuh *et al.* (2013), increasing productivity is at the core of every industrial revolution. The first three industrial revolutions had a strong impact on industrial processes, allowing to increase productivity and efficiency through the use of disruptive technological developments, such as steam engines, electricity or digital technology.

In recent years, along with increased research attention on the Internet of Things (IoT) and Cyber-Physical Systems (CPS) (KHAITAN; MCCALLEY, 2015), governments and industries around the world have noticed this trend, and acted to benefit from what this new wave of industrial revolution could provide (RIDGWAY *et al.*, 2013; SIEMIENIUCH *et al.,* 2015). "Industry 4.0" was first coined at the Hanover Fair in 2011, and the term drew much attention from academics, practitioners, government officials and politicians around the world (SUNG, 2018).

Industry 4.0, which may eventually represent a fourth industrial revolution, is a complex technological system, widely discussed and researched, with great influence in the industrial sector. Industry 4.0 is a generic term for a new industrial paradigm that encompasses a set of future industrial developments related technologies such as: *Cyber-Physical Systems* (CPS), Internet of Things (IoT), Internet of Services (IoS), Robotics, *Big Data*, Cloud Manufacturing and Augmented Reality. The adoption of these technologies is fundamental for the development of smart manufacturing processes, which include devices, machines, production modules and products able to independently exchange information, trigger actions and control themselves (WEYER *et al.,* 2015).

Figure 3 shows the evolution from Industry 1.0 to Industry 4.0, over industrial integration and industrial information integration for the emerging era of IoT and CPS (XU *et al.*, 2018).

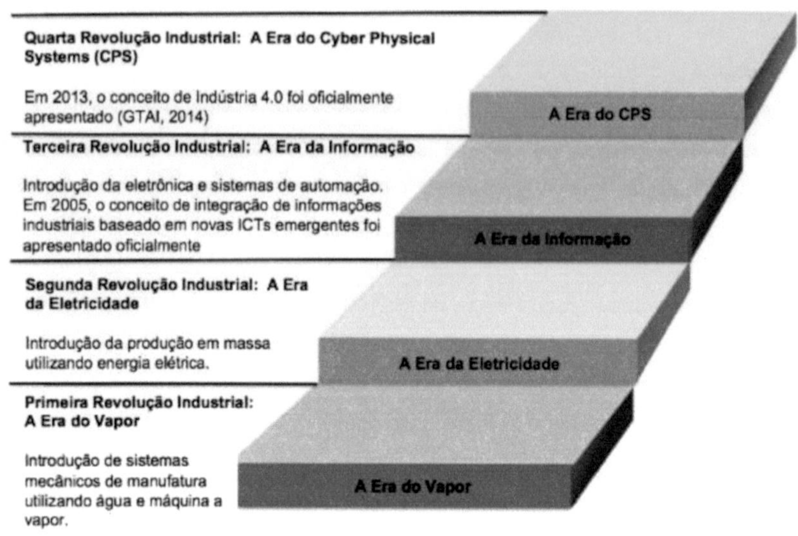

Figure 3 - The evolution from Industry 1.0 to Industry 4.0
Source: Xu *et al.* (2018)

According to Kagermann et al. (2013), Industry 4.0 takes manufacturing automation to a new level by introducing customised and flexible mass production technologies. This means that machines can operate independently or coordinate with humans to produce customer-oriented manufacturing.

2.2 TERMS USED

Germany was the first to publicly refer to the digitisation of industry as "*Industrie 4.0*" in 2011 (LASI et al., 2014). The term was then expanded to the Anglo-Saxon world as "*Industry 4.0*", and was passed on to other countries with similar meanings and words. The United States, for example, focuses on the term "*Smart*

Manufacturing", as do Japan and Korea. *General Electric,* on the other hand, uses this idea under the name "*Industrial Internet*" (OZTEMEL; GURSEV, 2020). Finally, other similar terms are found in the literature. Chart 1 presents these terms that are considered synonyms for Industry 4.0 and are used in the literature.

Table. 1 - Synonyms for Industry 4.0

Terms	Authors
"Smart Manufacturing"	Davis et al. (2012); Wallace, Riddick (2013); Dais (2014); Ivezic et al. (2014); Choi et al. (2015); Davis et al. (2015); Bogle (2017); Feng et al.; Hedberg (2017); Kusiak (2017); Sharp et al. (2018)
"Fourth Industrial Revolution"	Gorecky et al. (2014); Yue et al, (2015); Wang et al. (2016); Park et al. (2017)
"Smart Factory" / "Smart Industry	Wang et al. (2016); Dias (2014); Tjahjono et al. (2017)
"Advanced Manufacturing	Davis et al. (2012); Esmaeilian et al. (2016); Tjahjono et al. (2017)
"Digital Manufacturing"	Byrne et al. (2016); Chong et al. (2018)
"Intelligent Manufacturing	Bogle (2017); Zhong et al. (2017)
"Factory of the Future (FoF)	Zangiacomi et al. (2018)
"Industrial Internet"	Wang et al. (2015)

Source: Own Authors (2021)

In many countries, including Korea, the fourth industrial revolution is a more widely used term than Industry 4.0, because the term "fourth industrial revolution" is more appealing and familiar (SUNG, 2018).

According to Schneider (2018), the terms *"Smart Factory"* and *"Factory of the Future"* are not related to the level of the industry as a whole, but to the factory itself. Given that networking, integration and communication are limited to components within the factory. To some extent, these terms fall short of describing the network underlying Industry 4.0. Terms like "*Smart Manufacturing"* or "*Production 4.0"* (which is hardly used in scientific contributions) explicitly focus on the application of SPC in manufacturing processes. In a way, this implies that these terms neglect the product network and the corresponding opportunities for data-driven services and new business models.

"Smart Industry" as well as *"Industrial Internet"* are more related to the industry level. Although these terms have not been prevalent in the literature and can be interpreted as target states of Industry 4.0, the latter differs mainly from Industry 4.0 in the way the range of applications is defined. (ARNOLD et al., 2016).

Like the term Industry 4.0, the other terms suffer from terminological shortcomings and imprecisions. In particular, there does not yet seem to be a consistent definition for any of these terms. Consequently, no conceptual clarity would be achieved by simply putting a different term into use (SCHNEIDER, 2018).

Also, according to Schneider's (2018) research, the awareness of the scientific community and the public is by far the highest for the term Industry 4.0. Although the term was initially invented and used mainly in Germany, it is safe to say that research has adopted the term - "also beyond the German-speaking area".

2.2.1 Definitions of Industry 4.0

In the literature, there are several definitions found for the term Industry 4.0. Trappey et al. (2016) defined Industry 4.0 as a general concept that allows manufacturing with the elements of tactical intelligence, using techniques and technologies such as Internet of Things, cloud computing and *big data*. According to Schmidt et al. (2015), from a technical point of view, Industry 4.0 can be described as the increasing digitisation and automation of the manufacturing environment, as well as the creation of a digital value chain to enable communication between products and their environment and business partners.

For Schumacher, Erol and Sihn (2016), Industry 4.0 refers to recent technological advances in which the internet and technologies of serve as the backbone of integrating physical objects, human actors, intelligent machines, product lines and processes across the organisation to form a new type of smart, networked and agile value chain.

According to Hermann et al. (2016), the concept Industry 4.0 can be understood as a collaborative term for technologies and concepts that span the organisation's entire value chain.

Weyer et al. (2015) states that this new industrial paradigm covers the development of intelligent environments capable of bringing the real and virtual worlds closer through the use of the CPS, integrating devices, machines, modules and production products, triggering actions and controlling themselves autonomously. The author also presents, a new important aspect for Industry 4.0: the man-machine interface and the emergence of new types of jobs.

In short, Industry 4.0 has enormous potential, impacting the entire value chain by optimising production processes, improving product quality, strengthening the relationship between all stakeholders and offering new business models and new ways of operating (FOIDL; FELDERER, 2016).

2.3 CHARACTERISATION OF I4.0

Researchers and companies hold different views on the concept and visions of Industry 4.0, but there is a consensus on the main aspects that address the future manufacturing vision (QIN et al., 2016). Based on Pereira and Romero (2017), the elements of value creation in Industry 4.0 can be characterised as follows:

(1) *Smart Factory:* is the result of several developments consisting of integration, digitalization and use of flexible structures and smart solutions. These manufacturing solutions enable the creation of a smart environment along the entire value chain, enabling the performance of flexible and adaptive processes (RADZIWON et al., 2014). A smart factory environment consists of a new real-time integrative intercommunication between all manufacturing resources, which increases manufacturing efficiency and enables meeting highly complex market requirements (KAGERMANN et al., 2013).

(2) *Smart products:* are integrated into the entire value chain as an active part of the systems, monitoring their own production stages through data storage, being able to request the necessary resources and control the production processes autonomously. In addition, smart products, as end products, should be aware of the parameters within which they should be used, providing information about their status throughout the life cycle (KAGERMANN et al., 2013).

(3) *Business models:* are being highly influenced by Industry 4.0, as this new manufacturing paradigm implies a new way of communicating along supply chains. Business modelling is changing in recent years due to new industrial and market requirements. There are many opportunities to optimise value creation processes and integration across the supply chain to achieve self-organisation, integration and real-time communication capabilities (QIN et al., 2016).

(4) Customers: they are a key factor in all business models and Industry 4.0 brings a set of advantages for them, improving communication along the value chain and enhancing the customer experience. The high level of integration and autonomous information exchange will allow changing requirements in real time (QIN et al., 2016).

(5) *People:* Human beings will be strongly affected by increased knowledge work

and task uncertainty (STOCK et al. 2018).

From a technical point of view, this new industrial paradigm can be described as the increased digitisation and automation of the manufacturing environment, in addition to an increase in communication made possible by the creation of a digital value chain (OESTERREICH; TEUTEBERG, 2016). Industry 4.0 aims to achieve a continuous flow of information and planning across the different levels and entities of industrial value creation. There are three types of integration possible in Industry 4.0: horizontal integration, vertical integration and end-to-end engineering integration (WANG et al., 2016).

Horizontal Integration refers to the exchange of information between companies and participants throughout the supply chain, connecting all modules of value creation (STOCK et al., 2018). Vertical Integration refers to the integration of these elements across departments and hierarchical levels of an organization, from product development to manufacturing, logistics and sales (PEREIRA; ROMERO, 2017). The aim is to resolve control in a hierarchical manner, providing immediate access to planning information from the equipment to the MES and ERP level. This allows, among other things, the possibility of self-control systems. End-to-end engineering integration, on the other hand, describes the connection of all phases of a product's life cycle; from raw material acquisition to manufacturing, product use and end of product life. It will offer new possibilities and business models by applying information from across the lifecycle (STOCK et al., 2018).

2.4 PROCESS INTEGRATION

The use of Industry 4.0 technologies supports a high level of process integration. This integration enables the creation of smart products and processes that help meet rapidly changing market demands in the form of increased functionality and more complexity (PERSSON, 2016). The following Industry 4.0 principles are a consequence of the successful deployment of Industry 4.0 technologies and process integrations (CARVALHO et al., 2018).

(1) *Interoperability.* It is defined as the ability to perform the same function, even when exchanging machines and equipment, even from different manufacturers. Interoperability provides a reliable environment, extending several networks in a

manufacturing system (QIN et al., 2016).

(2) *Decentralization.* It is defined as the ability of companies, operations staff and even machines to make decisions. This principle helps to get quick decisions and offers more flexibility. Decentralization acts as a perfect organizational setup to deal with the growing needs of highly customized products (KAMBLE et al. 2018).

(3) *Virtualisation.* It means creating a virtual copy of the physical world. Virtualization is used for process monitoring and machine-to-machine communication. These simulation based virtual models are linked to the sensor data. Virtualization helps to notify the system faults and advances the security provisions (KAMBLE et al. 2018).

(4) *Real-time working. Big data* technologies increase the real-time capability of organisations. Big data collected from factories, customers, and data received from suppliers when analysed in real time, changes the way decisions are made and its impact on the profitability of industry 4.0-based organisations (KAMBLE et al. 2018).

(5) *Modularity.* Refers to modular production systems that can be adapted by replacing and expanding individual modules in a much more comfortable way. System modularity offers capacity adjustments in situations of seasonal fluctuations or changes in product production requirements. Modularity also facilitates the simulation of various manufacturing processes such as product *design*, production planning, engineering and production services and production as individual processes and interconnect them later on, offering interchangeability (QIN et al., 2016).

(6) *Service orientation.* Process integration brings the meaning of service orientation, as all entities of the production system are interconnected, facilitating the creation of the product service system. The flexibility and agility achieved as a result of service orientation allows organisations to respond to market changes more quickly. This allows the various stakeholders of organisations to come together and partner to co-create value for consumers (KAMBLE et al. 2018).

2.5 BUILDING BLOCKS

Industry 4.0 is changing the way products are manufactured and also the way orders are placed. This new revolution is based on current technologies, but the way to use these technologies has been reorganized to provide superior coordinated services. According to Kumar and Nayyar (2020), the base of Industry 4.0 is built from four basic physical elements: Smart Sensors, Control Systems, Connectivity and Smart Factories. Figure 4 exemplifies this base, which we can call Building Blocks.

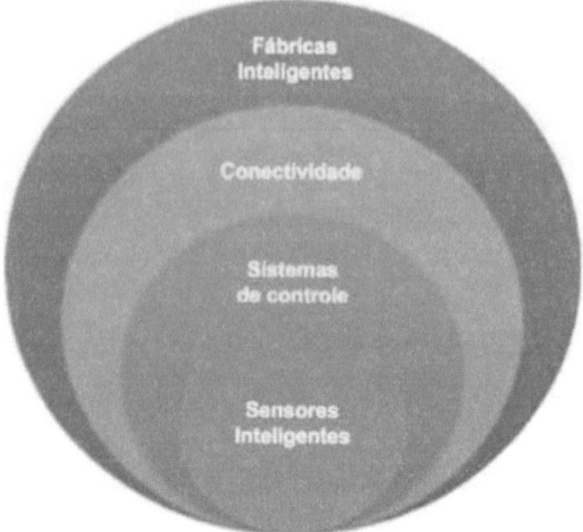

Figure 4 - Building Blocks of Industry 4.0
Source: Kumar and Nayyar (2020)

1) Smart sensors: they are the foundation of Industry 4.0. As I4.0 is a revolution involved in IoT and CPS, it is sensors that perform a wide range of functions to provide support for these technologies. Therefore, smart sensors are the first layer in the development of I4.0, as shown in Figure 4. Smart sensors act as manufacturing assets that can virtually gather big data about products and their environment. Sensors can provide wireless communication capability and data can also be synthesised using a cloud-based interface. Sensors can also detect anomalous activities (KUMAR; NAYYAR, 2020).

2) Control systems: monitor the working conditions of the manufacturing machines and any critical events. They serve as the brains of the manufacturing process. The control systems implemented in a manufacturing facility are usually centralised or decentralised. Subsystems like data acquisition systems, sensor network, actuators and control devices are highly involved in the control system for manufacturing.

Recently, control systems are involved in monitoring energy consumption in a manufacturing environment in real time (KUMAR; NAYYAR, 2020).

3) Connectivity: controls oversee the working conditions and the process of data acquisition by sensors, but this data will not be useful if it is not communicated to the centralized server or decision makers. This communication between devices is possible through connectivity and networking, and this connectivity is provided using the Internet. Several technologies are included: routers, servers, switches, *Ethernet*, *gateway* devices and *Linking* technologies. The basis of Industry 4.0 is the concept of everything connected; therefore, when a request is received by one device, all others are shared with the information by shared means. Mobile devices can play a key role in providing connectivity between devices. Communication is like the blood flow of industry 4.0 and telecommunications can play an important role in it (KUMAR; NAYYAR, 2020).

4) Smart Factory: The smart factory is the heart of Industry 4.0. The whole concept of business, products, machines and sales of the new revolution is part of or dependent on the smart factory. Smart factories have great control over themselves and provide a flexible system that allows self-optimisation of performance over a wide network and self-adaptation, enabling learning under real-time conditions. They also have the potential to autonomously run entire production processes. In simple terms, a smart factory is a factory in which smart machines work to produce smart products (KUMAR; NAYYAR, 2020).

Smart Sensors, Control Systems, Connectivity and Smart Factories are the need to configure a manufacturing industry for the current revolution. The technologies incorporated by this revolution are known as the nine pillars of Industry 4.0, which complete the *Building Blocks*. Figure 5 presents these technologies.

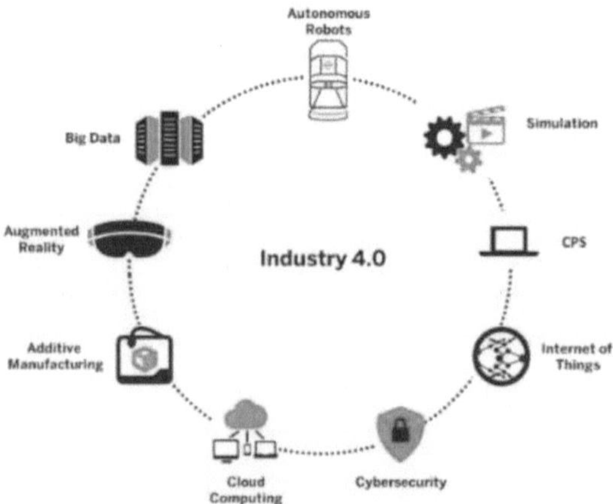

Figure 5 - Building Block Technologies
Source: Kumar and Nayyar (2020)

2.5.1 *Cyber and Physical System*

Cyber and Physical Systems (CPS) is the main foundation of Industry 4.0 (LEE et al., 2015). CPS are engineered systems that are built on and rely on the seamless integration of computational algorithms and physical components (XU et al., 2018). This allows objects to communicate with their environment and reconfigure in real time in response to new needs (YUE et al., 2015);

The CPS connects virtual space to physical reality, integrating computing, communication and storage resources; moreover, it can be real-time, efficient, reliable and safe (CHENG et al., 2016). CPS is considered an enabling technology for Industry 4.0 that will merge the virtual and physical worlds, making the boundaries between these two worlds disappear. Industry 4.0 manufacturing systems are collaborative systems that involve multiple communicating agents, including physical agents, *software* agents and human agents. This will result in the merging of technical and business processes, paving the way for a new industrial era, resulting in the smart factory. CPS can improve productivity and resource efficiency and enable more flexible models of work organisation (XU et al., 2018).

The essence of Industry 4.0 is to apply CPS to realise smart factories (KUSIAK, 2017). In other words, the smart factory is enabled by CPS-based production systems. CPS can play an important role in smart factory manufacturing and production processes. This provides significant advantages in real time, resources and costs compared to classical production systems (XU et al., 2018).

2.5.2 *Cloud Computing*

The operation of a modern company involves numerous decision-making activities, requiring a large amount of information and intensive computing. At one point, manufacturing companies required multiple computing resources, such as servers for databases and decision-making units. This caused inefficient data exchange and sharing, low productivity and less optimal use of manufacturing resources. Cloud computing provides an effective solution to these problems. All data can be stored on servers in public or private cloud and thus complex decision-making tasks can be supported by cloud computing (XU et al., 2018).

Cloud computing is a computing technology that offers high performance and low cost (ZHENG et al., 2014). Virtualization technology provides cloud computing with resource sharing, dynamic allocation, flexible extension, and numerous other advantages (XU et al., 2018). With reaction times of a few milliseconds and large bandwidths, sharing information across multiple systems and networks in real time can ensure that data and applications are available anywhere, all the time, and on any endpoint (GUPTA et al., 2013).

Cloud-based manufacturing is a growing technology that can significantly contribute to the realisation of Industry 4.0. It enables modularisation and service orientation in the manufacturing context, where system orchestration and sharing of services and components are important considerations (MOGHADDAM; NOF, 2017). Cloud manufacturing, similar to cloud computing, uses a network of resources in a highly distributed manner. Cloud design and manufacturing is considered the next paradigm in manufacturing, and extensive research is being conducted on its significance in Industry 4.0 (BRANGER; PANG, 2015).

2.5.3 *Industrial Internet of Things (IIoT)*

According to Wong and Kim (2017), the *Internet of Things (*IoT) is a new industrial ecosystem that combines intelligent and autonomous machines, advanced predictive analytics and human-machine collaboration to improve productivity and efficiency. It is a computing concept that describes the ubiquitous connection to the Internet, turning ordinary objects into connected devices (SISINNI et al., 2018). IoT provides real-time sensing / actuation capability and fast data / information transmission capability, so that remote operation of manufacturing activities and efficient collaboration among stakeholders are greatly facilitated (YANG et al., 2017).

As a subset of the IoT, the Industrial IoT (IIoT) covers the domains of machine-to-machine (M2M) communication and industrial communication technologies with automation applications. IIoT paves the way for a better understanding of the manufacturing process, enabling efficient and sustainable production. Smart manufacturing obviously focuses on the manufacturing stage of the life cycle of (smart) products, with the aim of responding quickly and dynamically to changes in demand. Therefore, IIoT affects the entire industrial value chain and is a requirement for smart manufacturing (SISINNI et al., 2018).

According to Xu, He and Li (2014), a typical IoT network includes four main essential layers:

(1) A sensing layer that integrates different types of "things" such as RFID tags, sensors, actuators;

(2) A network layer that supports information transfer over a wired or wireless network;

(3) A service layer that integrates services and applications through *middleware* technology; and

(4) An interface layer for displaying information to the user and enabling interaction with the system.

2.5.4 Augmented Reality (AR)

The availability of data in an embedded system provides new ways for users to access information. Smart glasses and other Augmented Reality (AR) and Virtual Reality (VR) technologies are increasingly being used in manufacturing processes. As such, they can be used to simulate an environment containing real objects, which can be used to enhance design and manufacturing processes (LEE et al., 2011).

Increasing human performance is the goal of AR, providing the necessary information for a specific task (PALMARINI et al., 2017). This new technology provides powerful tools, acting as a Human-Machine Interface (HMI). As a growing evolving technology, recently, the use of AR is spreading to different manufacturing fields. The use of AR in manufacturing processes related to simulation, assistance and guidance has proven to be an efficient technology that helps in problems. AR technology increases the operator's perception of reality using artificial information about the environment, where the real world is fulfilled by its objects (SYBERFELDT et al., 2015; SYBERFELDT et al., 2016).

Maintenance is one of the most promising fields of AR. It enhances human performance in performing technical maintenance tasks, as it also supports maintenance decision making (PALMARINI et al., 2017).

2.5.5 *Big Data Analytics*

Several tools and techniques are available to explore a large mass of production data. Processing large amounts of data has always been seen as a major challenge for production planning and control functions, as well as a necessity to achieve the goals of Industry 4.0 (BABICEANU; SEKER, 2016).

Big data analytics and technologies support real-time data collection from many different sources, leading to greater manufacturing flexibility, product quality, energy efficiency and enhanced equipment service (STRANGE; ZUCCHELLA, 2017; PREUVENEERS; ILIE-ZUDOR, 2017).

Big data technologies refer to a new generation of technologies and architectures that enable organizations to extract economic value through the discovery, capture and analysis of very large volumes of a wide variety of data. *Big data* analysis enables contemporary organizations to obtain better value from the enormous amount of information they already possess and identify what is likely to happen next and what actions should be taken to achieve the best results (LAVALLE et al., 2011).

Big data analytics has been widely used in manufacturing for process monitoring, and fault discovery supports new capabilities such as predictive analytics (YUAN et al., 2017). However, these features require high data quality and analytics expertise to realize efficient online manufacturing solutions (MOYNE; ISKANDAR, 2017). Exploiting *big data* intelligence to drive agility will require addressing new challenges, such as ensuring data consistency and confidentiality across long and complex supply chains (BOGLE, 2017).

2.5.6 *Cyber Security*

Ensuring a greater number of communication channels without reducing network performance is essential to ensure the deployment of Industry 4.0 strategies (AIREHROUR et al., 2016).

IoT-based PSCs, connected to millions of sensors and embedded communication devices, present the normal risks associated with increased data usage and the more significant risks of systemic breaches (WITTENBERG, 2016). CPMSs can face the threat of cyber attacks. Once CPMSs are deployed in organisations, malicious *software* can affect and spread from one machine to another through communication systems. The viruses spread by such *software* may be intended to modify the manufacturing process or destroy data through the system, leading to quality defects in products or a complete shutdown. Security issues in CPMSs in the form of cyber attacks and data theft are critical and need to be controlled to improve the reliability and acceptability of the system (YU et al., 2017). Industrial data is highly sensitive as it encompasses various aspects of industrial operation, including information about products, business strategies and companies (WOLF; SERPANOS, 2017).

Cybersecurity is a key element of Industry 4.0, given that all internet-facing organisations are at risk of attack. There is no doubt that Industry 4.0 will be challenged by traditional cybersecurity problems, along with its own security and privacy issues (THAMES; SCHAEFER, 2017).

2.5.7 Autonomous Robots

Robots and embedded sensor technologies are becoming increasingly flexible, communicative and cooperative (MICHNIEWICZ; REINHART, 2014). Product connectivity and appropriate mechanisms for collaboration with humans will eventually favour the reduction of batch sizes for a single item at a reasonable cost (MOEUF et al., 2018).

Modern robots are characterised as systems that offer autonomy, flexibility and cooperation. These robots offer cost advantage and an excellent range of features, performing most processes in the smart factory (PEI et al., 2017).

Processes such as product development, manufacturing and assembly phases are processes that adaptive robots are very useful in manufacturing systems (SALKIN et al., 2018). Importantly, fully autonomous robots make their own decisions to perform tacks in constantly changing environments without operator interaction (BEN-ARI; MONDADA, 2018).

The concept of collaborative robots (cobots) also introduces the closeness of robots to humans (KOCH et al., 2017). Cobots are a category of robots specially designed to interact directly and physically with humans in close cooperation. This is possible due to existing safety limits on speed and forces that reset the cobot automatically, allowing it to be guided manually (WEISS; HUBER, 2016). With this, for manufacturing companies, the human-robot barrier is broken, offering greater accessibility and flexibility in solutions (MAKRINI et al. 2018).

2.5.8 *Additive manufacturing / 3D Printing*

Additive Manufacturing is the process of joining materials to create three-dimensional objects from digital computer models using a machine. The process is usually developed layer after layer, unlike traditional subtractive manufacturing methods. The term 3D printing is much more popular and is commonly used as a synonym for *Additive Manufacturing Technologies* (AMT) (WOHLERS; CAFFREY, 2013).

Jian et al. (2017) discussed the potential of AMT in replacing many conventional manufacturing processes. AMT is an enabling technology that helps in new products, new business models and new supply chains. Unique products can be manufactured without conventional surplus, so it is a great advantage. With 3D printing it is possible to create prototypes to enable independence from elements of the value chain and therefore achieve time reduction in the design and manufacturing process.

Although 3D manufacturing systems are still in the early stages, organisations are expected to make much wider use of it in Industry 4.0. Chen and Lin (2017) assert the need for the use of 3D printing technology as an enabler for smart manufacturing. However, they identify that 3D printing faces technical challenges such as limited types of usable materials, low precision and low productivity. If these challenges are overcome, smart manufacturing based on this technology could continuously deliver 3D objects of interest to customers or join as many ubiquitous production networks as possible. Zawadzki and Zywicki (2016) discuss hybrid prototyping, which is an integration of additive manufacturing technologies and virtual reality techniques.

2.5.9 Simulations

Simulations and prototyping will leverage real-time data to mirror the physical world in a virtual model, which can include machines, products, and humans (RÜßMANN et al., 2015). This allows operators to test and optimise machine settings for the next in-line product in the virtual world before the physical changeover, thus decreasing machine setup times and increasing quality (BAHRIN et al., 2016).

Simulation and modelling techniques aim at simplifying and economically favouring the design, realization, testing and execution of an active operation of manufacturing systems (KOCIAN et al., 2012). Simulation and modelling not only allow manufacturers to avoid mistakes at an early stage that could otherwise result in substantial costs for plant operators, but can also be used to optimise a manufacturing plant during ongoing daily operation (GILCHRIST, 2016).

Optimal or near-optimal system design is the goal of decision makers. This optimization is possible due to a systematic search in a wide decision space, without pre-specified constraints or requirements. In dynamic and uncertain environments, this tool has the potential to optimize control decisions and support real-time decision making. This can be possible when the required computational efficiency is achieved. Compared to conventional simulation, online real-time simulation can analyse user and system behaviour in milliseconds, allowing the user to practically develop and produce a prototype for the product or service (XU et al., 2016). According to Cedeño et al. (2018), a real-time simulation is when a computer runs at the same rate as the physical system; therefore, the simulation model needs to be fed with real-time data that can be achieved using IoT.

Together, these nine building blocks drive the interconnection of all sensors, IT systems and machines across the enterprise and its supply chain.

2.6 IMPACTS AND BARRIERS

Advances in digital transformation and increasing interconnectivity will bring new challenges for organisations as Industry 4.0 will significantly change products and manufacturing systems in design, processes, operations and services (PEREIRA; ROMERO, 2017).

Industry 4.0 will lead to possible profound changes in several domains that go beyond the industrial sector. However, all this change in the industrial landscape brings a number of barriers that hinder its implementation.

According to Pereira and Romero (2017), its impacts and influence can be classified into six main areas: (1) Industry, (2) Products and services, (3) Business models and market, (4) Economy, (5) Work environment and (6) Skills development. Thus, all barriers found in the literature review were distributed among these six main areas. Table 2 presents these barriers:

Table. 2 - Industry 4.0 barriers

Areas	Barrier	Authors
Industry	Infrastructure	Thoben et al. (2017)
	Lack of security and privacy / IT	Thoben et al. (2017); Sung (2018); Marques et al. (2017); Hecklau et al. (2016); Yu et al. (2017); Babiceanu and Seker (2016)
	Complexity regarding the total integration of products and processes	Dombrowski, Wagner (2014)
	Organisational and process changes	Bröring et al. (2017); Gnimpieba et al. (2015); Hussain (2017); Valmohammadi (2016)
	Lack of internet coverage and IT facilities	Fang et al. (2016); Yan et al. (2014a); Yan et al. (2014b).
	Need to maintain the integrity of production processes	Sung (2018).
	Regulatory compliance issues	Lin (2016); Thoben et al. (2017); Christian et al. (2017)
	Legal and contractual uncertainty	Christians, Liepin (2017); Shelbourn et al. (2005); Merz, Mangini (2002); Thoben et al. (2017) Christian et al. (2017); Marques et al. (2017).
Products and services	Growing demand for the development of more complex and intelligent products	Davis et al. (2012)
	Customer demand	Thoben et al. (2017)
	Technological immaturity	Marques et al. (2017); Thoben et al. (2017)
Business models and market	Very rapid change in business models	Thoben et al. (2017)
	New competitors	Thoben et al. (2017)
	Integration and compatibility issues	Valmohammadi (2016); Hussain (2017); Da Xu et al. (2014)
Economy	Investment costs	Thoben et al. (2017)
	High implementation cost	Kamigaki (2017)
	Uncertain return on investment	Marques et al. (2017);

Work environment	Loss of many jobs	Sung (2018); Frey, Osborne (2017); Spath et al. (2013)
	Reliability and stability required for critical machine-to-machine (M2M) communication	Sung (2018); Kagermann et al. (2013)
	Man-Machine Relationship	Rauch et al. (2019); Santana et al. (2017)
	General reluctance to change by stakeholders	Sung (2018).
Skills development	Lack of knowledge	Christian et al. (2017); Lorna (2017); Davies et al. (2011); Rymaszewska, Helo (2017)
	Lack of required skills of employees	Thoben et al. (2017); Sung (2018); Koch et al. (2014); Mueller et al. (2017) Christian et al. (2017); Marques et al. (2017); Benešová, Tupa (2017); Ryan et al. (2017)
	Need for high quality training	Erol et al. (2016)
	Interdisciplinary thinking	Magruk (2016)
	Need for increasingly qualified personnel in technological areas	Pereira, Romero (2017)
	Lack of clear understanding about the benefits of IoT	Da Xu et al. (2014); Hognelid, Kalling (2015); Lee, Lee (2015)
	Lack of standards and reference architecture	Mueller et al. (2017)

Source: Own Authors (2021)

Perhaps the most challenging aspect of implementing Industry 4.0 is the IT security risk. Industry 4.0 will require online integration between multiple entities, and this online integration will make room for security breaches and data leaks. Cyber theft would be another dangerous threat. In this case, the problem is not individual and this will cost manufacturers substantially and may even damage their reputation. Therefore, security is a crucial issue that should be treated seriously (SUNG, 2018).

The transformation to Industry 4.0 will require major investments in new technologies, and the decision for these transformations should be made at the CEO level. Even then, the risks must be calculated and taken seriously. While it is too early to speculate on employment issues with the advent of Industry 4.0, it is safe to accept that workers will need to acquire different or entirely new skills. This may help increase employment rates, but it will also alienate a large sector of workers (SUNG, 2018).

The sector of workers whose work is perhaps repetitive and routine will face a strong challenge to keep their jobs. New and different educational systems should be introduced, but this still does not solve the problem of older workers. This is a problem that may take a long time to solve. Privacy is not only a concern for the customer, but also for the manufacturer. In an interconnected Industry 4.0 network, manufacturers need to collect and analyse a huge amount of data. For customers, this can seem like a threat to their privacy. Reducing the gap between consumer and manufacturer will be a big challenge for both parties (SUNG, 2018).

3. TECHNOLOGY TRANSFER

To better contextualize technology transfer in this paper, it is first necessary to understand the concepts that surround the theme. The term *technology* focuses on the knowledge of a specific technique and method to solve a problem. Technology has been developed by scientific research and R&D, and is a critical element for the economic development of industry. By improving the efficiency of a firm's activities, technology helps reduce the cost of production and increase manufacturing productivity (GISSELQUIST; GRETHER, 2000).

For Kahn et al. (2005), technology is a process whereby knowledge, innovations and discoveries are acquired and developed. Subsequently, this acquired knowledge is used in product projects. Figure 6 presents this idea:

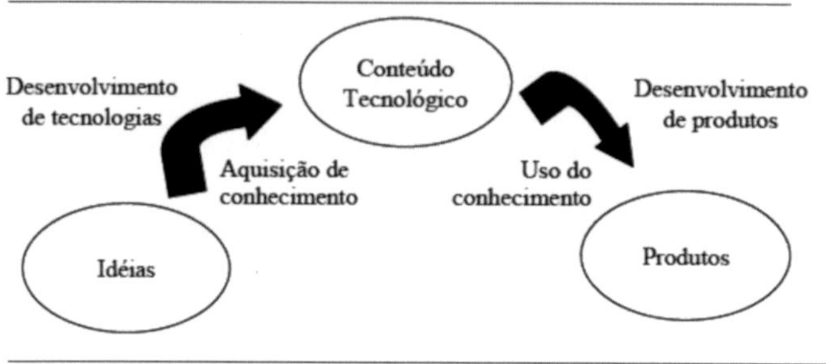

Figure 6 - Technology Development
Source: Kahn et al. (2005)

Technology can be understood as knowledge. The term Knowledge has been used in organizations since 1991. However, only in 2004 that this term started to be recognized as an organizational performance (ANH; CHANG, 2004). Nowadays, researchers already adopt the term "Knowledge Management System" (HUANG et al., 2014).

The moment the company needs a new technology, it has options to follow. The first is to use and develop knowledge through its own resources. The second option is to acquire knowledge and technology from third parties, which have greater opportunities for a better result, as expected, because they already have the necessary knowledge. This second option is the process we call Technology Transfer (TT).

TT is an important tool, which allows a company to improve its competitive advantage, its financial and technological benefits. Lee *et al.* (2010) points out the benefits of technology transfer in production processes:

- Improved income process;
- Improvement in design products and services;
- Improved design for the market;
- Standardisation;
- Physical product properties and performance characteristics;
- Ability to switch from intermittent to mass flow processes.

3.1 DEFINITIONS OF TECHNOLOGY AND KNOWLEDGE TRANSFER

Schnepp et al. (1990) define technology transfer as the process by which experience or knowledge related to some aspect of technology is passed from one user to another, with the aim of economic gain.

According to Autio and Laamanen (1995), technology transfer is an intentional, goal-oriented interaction between two or more social entities, during which the body of technological knowledge remains stable or increases through the transfer of one or more technology components.

For Bozeman (2000), the definition of technology transfer may coincide with the definition of production transfer. Therefore, it is not always easy to distinguish the two concepts.

Bozeman (2000) further defines TT as the process that allows the passage of a technology from one organisation (donor) to another entrepreneurial organisation (receiver).

The process of technology transfer is presented by a range of literature on innovation, development, behaviour change and economic development (KARAKOSTA *et al.,* 2010).

According to Debackere et al. (2014) and Osabutey and Jin (2016), TT is defined as the process of transfer or dissemination of technology from its origin to a wider distribution, to more people and places. It occurs on several axes: between universities, from universities to companies, from large companies to smaller ones, from governments to companies, across borders, formally and informally.

For Ismail, Hamzah and Bebenroth (2018), Technology Transfer is the process where technologies are distributed from their place of origin to more people and places, being influenced by information characteristics.

For the definitions given, one can perceive that in all of them, technology transfer involves the acquisition of innovation from an external source as well as the sharing of technological knowledge of their products or processes. It is also perceived that Technology Transfer can be a knowledge transfer.

Argote and Ingram (2000) define Knowledge Transfer as "the process by which a unit (e.g., group, department or division) is affected by the experience of another". It also refers to a dyadic exchange between individuals, groups or organizations, in which a recipient can understand, learn and apply knowledge transmitted from a source (HAMID; SALIM, 2011; ISMAIL, 2015).

3.2 THE TT PROCESS

Technology transfer is often referred to as licensing, a technology cooperation. Technology transfer is not only a simple way of technology trading, but also can be a source of useful information. Through technology transfer, the firm can improve its market share and its current state of technology (PARK; LEE, 2011).

The basic concept of the TT process is the movement of technology from one place to another, whether in the form of products, knowledge or technology. According to Cormican and O'Connor (2009), TT can occur in the different ways:

- Transfer from universities and research laboratories to industrial companies;
- Transfer within an organisation, from research to commercialisation;
- Transfer from one organisation to another;
- Transfer between countries;
- Transfer through sale of products incorporating the technology;
- Transfer through contractual arrangements, including licensing, cooperation and sharing between companies as part of strategic alliances;
- Transfer of tangible assets, such as new products, facilities and equipment;
- Transferring intangible forms through formal mechanisms such as patents and licences;
- Informal transfer through knowledge and information flows.

Technology transfer has three different types of technology cooperation. If a technology or innovation is acquired from an external source, it is referred to as *licensing-in*. If the knowledge or technology is sold or donated to other companies it is referred to as *licensing-out*. Technology cooperation is not only in one direction, but a bi-directional technology transfer (PARK; LEE, 2011).

An important distinction in the TT literature is between vertical and horizontal technology transfer. Both are considered important for economic development, and can be present together in an organisation. In horizontal technology transfer, knowledge is diffused to competitors, while in vertical technology transfer, knowledge is diffused to potential input suppliers (OCKWELL *et al.*, 2008).

- Vertical technology transfer: is the transfer of technological knowledge or innovation, from basic to advanced research, from the R&D phase to product commercialization (OCKWELL *et al.*, 2008; PARK; LEE, 2011).

- Horizontal technology transfer: is the transfer of technological knowledge or innovation between projects, organizations, industries and nations. It is considered as the transfer carried out from a geographical location to another (OCKWELL *et al.*, 2008; PARK; LEE, 2011).

According to Mendoza and Sanchez (2018), there are three main agents in a technology transfer. These being:

- The technology supplier: also known as supplier, generator or transmitter. They are usually universities, research organizations, technology centres, companies with research capacity (R&D) and knowledge generation.

- The technology receiver: also called user, client, beneficiary or assignee, is the entity that buys or benefits from the technology transfer. These are usually companies with technological innovation projects.

- Intermediary: known as an accelerator, facilitator, consultant or diffuser, are the people (consultants, lawyers) or supporting institutions (development agencies, foundations, associations, chambers of commerce, institutional marketing structures, institutional marketing structures, technology transfer offices (TTOs), offices with commercial links) of public or private character that support TT providers or recipients.

Another important point in the TT process, discussed by several authors (OCKWELL *et al.*, 2008; SHUJING, 2012; PUEYO *et al.*, 2011) are the three different flows of technology transfer, from the lowest to the highest impact on the technological capabilities of the recipient. Figure 7 presents this flow:

Figure 7 - Technological content of TT
Source: adapted from Shujing (2012)

The first flow includes capital goods and equipment. This flow increases the production capacity of the recipient, but does not allow the recipient to use imported equipment efficiently or generate technological change. The second flow includes skills and *know-how* for the operation and maintenance of equipment. It brings the human resources of the importer to the technological level needed to efficiently operate the imported technology, but without efforts, it does not allow technological change. The third stream encompasses the knowledge and expertise to generate and manage technological change. It creates new technological capability through the recipient's knowledge transfer, creation and innovation (OCKWELL et al., 2008; SHUJING, 2012; PUEYO et al., 2011).

3.3 UNIVERSITY-INDUSTRY INTERACTION

One of the main ways of transferring technology is through the interaction of universities with industries.

The role of universities in the knowledge society as producers and transmitters of knowledge is increasingly important, not only as instructors of human capital, but also in driving and creating inventions and innovations and solving social problems (DUTRÉNIT et al., 2010). This enables the use, transfer and commercialisation of knowledge as part of a new mission (SIERRA; VILLAZUL, 2018).

Based on Debackere and Veulegers (2005), university-industry collaboration refers to different types of interactions between industry and the scientific sector, which aim to exchange technology. The interaction channels involve hiring recent graduates, conferences, staff training, patents, prototypes, licensing, incubators, spin-offs, joint and cooperative ventures. However, the most common are the traditional channels, involving publications, conferences and consultancy (BARLETTA et al., 2017; LEMOS; FERRAZ, 2017; DE FUENTES; DUTRÉNIT, 2012; TORRES et al., 2011; ARVANITIS; KUBLI; WOERTER, 2008; ARVANITIS; SYDOW; WOERTER, 2008; BEKKERS; BODAS, 2008; ARZA, 2010; DUTRÉNIT; ARZA, 2010).

One of the main methods for generating this knowledge transfer is university *spin-offs*. In the commercial field, the term spin-off refers to the process by which a company is created from another pre-existing entity. The resulting new company is also known as a *spin-off*. The term serves to define the process itself and the outcome of that process. At the university level, one speaks of an academic or university *spin-off* when the company was established within a higher education institution, putting into practice the knowledge generated at these centres through the R&D activity of academics (MIRANDA et al., 2018).

Etzkowitz and Leydesdorff (2000), developed a model where collaboration between three institutional spheres (government, academia and industry) is considered of critical importance to enhance regional economic and social development.

The inclusion of the fourth helix, which represents users of society-based innovation (LEYDESDORFF, 2011; CARAYANNIS et al., 2012) and the subsequent emergence of quadruple helix structures reflect the review by Bozeman et al. (2015), demonstrating their growing importance. Figure 8 presents the transition from triple helix to quadruple helix structures.

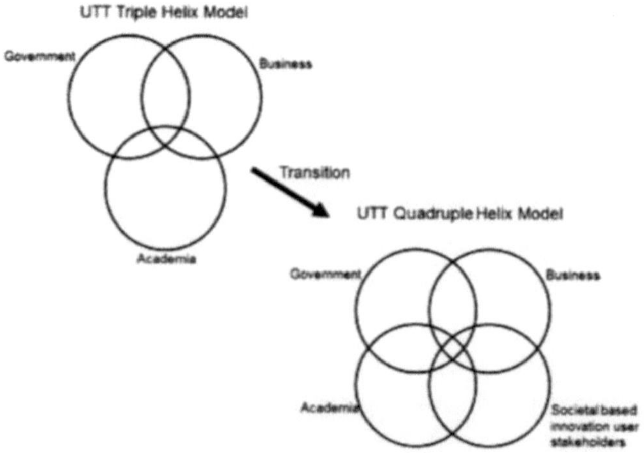

Figure 8 - Transition from triple helix to quadruple helix structures
Source: Miller et al. (2018)

According to Miller et al. (2018), over the past three decades, collaboration between university, government and industry has evolved considerably. This evolution is partly due to a combination of globalisation and regionalisation in economic development (MCADAM et al., 2012) with increasing government pressure on universities to take a more proactive role in regional and social development (GRIMALDI et al., 2011).

The inclusion of the fourth helix (Society-based innovation) reflects the development and increasing complexity and change of modern economic systems, which suggests that the triple helix is not sufficient to ensure long-term sustainable growth (MACGREGOR et al., 2010; IVANOVA, 2014).

According to Agrawal (2001), it is commonly accepted that universities are an important source of new knowledge, especially in the areas of science and technology. Therefore, it is important to build as clear a picture as possible of the mechanisms by which university science moves into the economy.

3.4 TECHNOLOGY TRANSFER CHANNELS AND MECHANISMS

Technology transfer is carried out through certain mechanisms. Karakosta et al. (2010) indicate that through these mechanisms there is a transfer of skills, knowledge and equipment mastery.

Autio and Laamanen (1995) provide a definition of a transfer mechanism: "a technology transfer mechanism is any specific form of interaction between two or more social entities during which technology is transferred". When interaction is continuous between the parties, a stable connection is established through which knowledge and technology flow.

To rationalise the large number of different mechanisms, channels or modes presented in the literature, the classification proposed by Autio and Laamanen (1995) was used. The authors divide the transfer mechanisms into three categories:

- Process mechanisms (service);
- Process mechanisms (organisational modes);
- Output mechanisms (search results).

Table 3 presents the technology transfer mechanisms found in the systematic literature review, divided into these three categories.

Table. 3 - Technology Transfer Mechanisms

TT MECHANISMS	AUTHORS
PROCESS MECHANISMS (SERVICE)	
Training; Human Resources Development	Bozeman (2000); Davenport and Prusak (2000); Szulanski (2000); Stephan (2001); Amesse and Cohendet (2001); Caputo et al. (2002); Argote et al. (2003); Cummings and Teng (2003); Lee, Win (2004); Huanca (2004); Siegel et al. (2004); Debackere, Veugelers (2005); Leloglu, Kocaoglan (2008); Hussler, Picard, Tang (2010); van der Gaast et al. (2009); Karakosta et al. (2010a); Flamos and Begg (2010); Karakosta et al. (2010b); Jason et al. *(2010);* Azam (2011); Baraki and Brent (2013); Blohmke (2014); Urban et al. *(*2015); Adenle et all. (2015); Corradini et *al.* (2016); Mulamula and Amadi-Echendu (2017); Verdolini and Bosetti (2017); Khan et al. (2017); George *et al. (*2017); Chege et al. *(*2019); Morgera and Ntona (2018); Lema et al. *(*2018); Eitzel et al. (2018); Bruckman *et al. (2018).*
Transfer of knowledge and *know how*	Siegel et al. (2004); Gerstlberger (2004); Ravindranath and Balachandra (2009); Lybæk and Andersen (2010); Flamos and Begg (2010); Parnphumeesup and Kerr (2011Pinard (2013); Baraki and Brent (2013); Blohmke (2014); Urban *et al.* (2015); Mulamula and Amadi-Echendu (2017); Aggarwal and Aggarwal (2017); Basu (2018); Chege et *al. (*2019); Morgera and Ntona (2018); Li *et al. (2018);* Bliznets et al. (2018).
Exchange of staff,	Lee, Win (2004); Khalozadeh et al., (2011); Popp (2011); Eaton (2013); Makarewicz- Marcinkiewicz (2013); Blohmke (2014); Ignatavičius *et al. (*2015);

researchers or professionals	Verdolini and Bosetti (2017); Manyuchi (2017); Basu (2018); Bliznets et al. (2018); Lema *et al. (2018)*; Sarkodie and Strezov (2019).
Consultancies	Bozeman (2000); Stephan (2001); Caputo et al. (2002); Lee, Win (2004); Huanca (2004); Siegel et al. (2004); Debackere, Veugelers (2005); Khalozadeh et al., (2011); Trencher et al. *(*2013); Vac and Fitiu (2017); Bliznets et al. (2018) .
Research projects in cooperation between universities and industry	Bozeman (2000); Amesse, Cohendet (2001); Lee, Win (2004); Huanca (2004); Stephan (2001); Autio et al. (2004); Siegel et al. (2004); Debackere, Veugelers (2005); Looy et al. (2011).
Meetings to exchange information	Cohen et al. (2002); Lee, Win (2004); Debackere and Veugelers (2005); Leloglu, Kocaoglan (2008); Hussler et al. (2010).
Laboratory demonstrations	Bozeman (2000); Canestrino (2009).
Company visits and presentations	Bozeman (2000); Canestrino (2009).
Networking and resource sharing events	Bozeman (2000); Canestrino (2009).
Internship	Cohen et al. (2002).
Transfer of information (via manuals, reports etc.)	Malik (2002).
PROCESS MECHANISMS (ORGANISATIONAL MODE)	
Inter-institutional networks; Industrial symbiosis	Bozeman (2000); Argote et al. (2003); Cummings, Teng (2003); Albors et al. (2005); Spalding-Fecher et *al. (*2005); van der Gaast et al. *(*2009); Lybæk, Andersen (2010); Flamos, Begg (2010); Karakosta et al. (2010a); Karakosta et al. (2010b); Gilsing et al. (2011); Kang, Park (2013Agboola (2014); Kruckenberg (2015a); Kruckenberg (2015b); Urban et *al. (*2015); Tvaronavičienė, Černevičiūtė(2015); Mehta *et al. (*2016); Morgera, Ntona (2018); Bruckman et al. (2018); Li *et* al. (2018); Sarkodie, Strezov (2019)
Research and Development	Stock, Tatikonda (2000); Amesse, Cohendet (2001); Lee, Win (2004); Salicrup ,Fedorkova (2006); Karakosta, Doukas, Psarras (2010); Murthi ,Shoba (2010); Sugandhavanija et al. (2011); Liu et *al. (*2013); Baraki and Brent (2013); Pinard (2013); Urban et al. *(*2015); Mehta et al. *(*2016); Aggarwal and Aggarwal (2017); Li *et al. (*2018); Bliznets et al. (2018)
Collaborations with customers or suppliers on joint development activities	Bozeman (2000); Reisman (2005); Stock and Tatikonda (2000); Cummings and Teng (2003); Canestrino (2009); Ravindranath and Balachandra (2009); Lybæk and Andersen (2010); Karakosta et al. (2010a); Karakosta et al. (2010b); Agboola (2014); Kruckenberg (2015a); Kruckenberg (2015b); Urban et *al. (2015*); Tvaronavičienė and Černevičiūtė (2015); Mehta *et al. (*2016); Morgera and Ntona (2018);
Spin-offs	Bozeman (2000); Rogers et al. (2000); Etzkowitz; Leydesdorff (2000); Autio et al. (2004); Siegel et al. (2004); Canestrino (2009); Hussler et al. (2010); Murthi, Shoba (2010); Koumpis et al. (2010); Gilsing et al. (2011); Looy et al. (2011); Sánchez et al. (2012); Trencher *et al. (*2013); Vac, Fitiu (2017)
Technology parks	Hussler, Picard, Tang (2010); Lee, Win (2004); Petroni et al. (2013); Huanca (2004); Siegel et al. (2004); Stephan (2001); Agboola (2014); Kruckenberg (2015a); Kruckenberg (2015b); Urban *et al.* (2015); Ignatavičius *et al. (2015*);

	Tvaronavičienė and Černevičiūtė (2015); Morgera and Ntona (2018).
Joint Venture	Stephan (2001); Huanca (2004); Siegel et al. (2004); Schneider et al. (2008); Mohamed et al. (2010); Khalozadeh et al. (2011); Mohamed et al. (2012); Urban *et al.* (2015); Mehta *et al.* (2016); Bliznets et al. (2018)
Buying ready-made technologies	Stephan (2001); Lee, Win (2004); Huanca (2004); Siegel et al. (2004); Lovett et al. (2009); Flamos and Begg (2010); Mehta *et al.* (2016); Bliznets et al. (2018)
Benchmarking	Cohen et al. (2002); Debackere and Veugelers (2005); Leloglu, Kocaoglan (2008); Rasmussen, Rice (2011).
Strategic alliances	Amesse, Cohendet (2001); Cummings, Teng (2003).
OUTPUT MECHANISMS (SEARCH RESULTS)	
Intellectual Property Law	Bozeman (2000); Amesse, Cohendet (2001); Caputo et al. (2002); Lichtenthaler, Ernst (2007); Karakosta, Psarras (2009); Lovett *et al.* (2009); Flamos and Begg (2010); Karakosta, Doukas and Psarras (2010a); Azam (2011); Eaton (2013); Blohmke (2014); Tvaronavičienė and Černevičiūtė (2015); Adenle et al. (2015); Mehta et al. (2016); Verdolini and Bosetti (2017); Basu (2018); Bliznets et al. (2018)
Licensing	Bozeman (2000); Fosfuri (2000); Lee, Win (2004); Debackere, Veugelers (2005); Schneider et al. (2008); Hussler et al. (2010); Sun et al. (2013); Silva et al., (2013) ; Bozeman et al. (2015); Leloglu, Kocaoglan (2008); Eaton (2013); Makarewicz-Marcinkiewicz (2013); Mehta et al. (2016); Aggarwal and Aggarwal (2017); Vac and Fitiu (2017); Bliznets et al. (2018) .
Publications	Davenport and Prusak (2000); Bozeman (2000); Amesse and Cohendet (2001); Cummings and Teng (2003); Caputo et al. (2002); Ferdows (2006); Gilsing et al. (2011); Trencher et al. (2013); Pinard (2013); Escalante et al. (2013).
Seminars; Conferences	Davenport, Prusak (2000); Bozeman (2000); Amesse, Cohendet (2001); Caputo et al. (2002); Cummings, Teng (2003); Ferdows (2006); Trencher et al. (2013); Pinard (2013); Belmonte *et al.* (2015); Eitzel et *al.* (2018).
Patents	Gomes-Casseresa et al. (2006); Hussler, Picard, Tang (2010); Gilsing et al. (2011); Looy et al. (2011); Park et al. (2013); Trencher et al. *(2013);* Shi and Lai (2013); Bliznets et al. (2018); Ferreira et *al. (2018).*
Demonstration of successful projects	van der Gaast *et al.* (2009); Ravindranath and Balachandra (2009); Karakosta et al. (2010a); Karakosta et al.(2010b); Pearce et *al.* (2012);
Continuing education programmes	Hussler et al. (2010).

Source: Own Authors (2021)

There are formal and informal TT mechanisms. Formal mechanisms are those that involve a legal instrument, such as patents, licences or *royalty* agreements. An informal technology transfer mechanism facilitates the flow of knowledge, but through an informal communication process, such as technical assistance, consultancy and joint research (LINK et al., 2007).

While formal mechanisms are appropriate to capture and transfer the explicit part of the technology, other approaches are needed to share the tacit component, which is of a non-codifiable nature. By simply studying equipment operation manuals and other written documentation, a buyer cannot capture the true capabilities of the equipment. Tacit knowledge has to be transferred through human interaction, such as first-hand observation. Therefore, a good buyer-supplier relationship and knowledge management are necessary for the technology transfer of high-tech equipment (LEE *et al.,* 2010).

There is also the difference between the unidirectional mechanism (based on the dissemination of research results) and the bidirectional mechanisms, which involve interactive development and service. In addition to this preliminary division, process mechanisms are divided into "services" and "organizational modes" to emphasize the difference between activities performed in existing units and the creation of new and appropriate organizational structures (BATTISTELLA et al., 2016).

3.5 BARRIERS TO TECHNOLOGY TRANSFER

Since technology transfer is one of the most complex learning processes, the effective transfer may not be possible until all factors (useful factors known as "facilitators" and hindrances called "barriers") related to this process are well explored and understood (SINGH; ABHISHEK, 2013). Barriers to technology transfer can also be seen as opportunities for intervention for technologies to reach their full potential (SATHAYE et al., 2001). Good technology transfer can enable an organization to improve manufacturing productivity, alliance efficiency and adaptability, international expansion and sustainable competitive advantage (LEE et al., 2010).

These barriers can be classified into several categories. For example, Riege (2005) and Kukko (2013) grouped the barriers to an organisation's growth into three categories: individuals, organisations and technology. Patil and Kant (2014) divided barriers into five categories: strategy, organisation, technology, culture and people. Nidhra et al. (2013) classified Knowledge Transfer barriers in relation to global software development into three categories: people, projects and technology. According to Greiner and Franza (2003), these barriers are basically divided into three groups: technical, regulatory or policy, and personal or institutional barriers. This study applied the categories developed by Greiner and Franza (2003).

After the study of bibliometry in this present work, all the barriers found in the articles, which present themselves in a relevant way, were extracted and divided into the three groups proposed by Greiner and Franza (2003). Table 4 presents these barriers found in the TT.

Table. 4 - Barriers to Technology Transfer

	BARREIRAS	AUTHOR
Technical Barriers	Intellectual property law	Dasgupta and Taneja (2011); Azam (2011); Karakosta et al. (2012); Correa (2013); Rai et al. (2014); Morgera and Ntona (2018)
	Risk of conflict of interest	Gilsing et al. (2011); Van der Gaast et al. (2009); Karakosta et al. (2010a); Flamos et al. (2010); Karakosta et al. (2011); Parnphumeesup, Kerr (2011); Pearce *et al.* (2012); Khan et al. (2017)
	Technical risk of the new technology	Pérez, Sánchez (2003); Karakosta et al. (2009); Karakosta, Psarras (2009); Ravindranath and Balachandra (2009); Silva Jr. *et al.* (2013); Khan, Haleem and Husain (2017); Chege *et al.* (2019)
	Feature of the technology	Koefoed, Buckley (2008); Gallagher (2006); Flamos *et al.* (2010); Flamos (2010); Doukas et al. (2012); Torvanger et al. (2013); Blohmke (2014); Belmonte *et al.* (2015); Khan et al. (2017)
	Scientific knowledge being too general to be useful to the company	Gilsing et al. (2011); Ravindranath and Balachandra (2009); Karakosta et al. (2010a); Baraki and Brent (2013); Belman and Tzachor (2013); Vac and Fitiu (2017); Chege *et al.* (2019); Morgera and Ntona (2018)
	Poor market infrastructure	Kennedy, Basu (2013); Ravindranath and Balachandra (2009); Jason *et al.* (2010); Adenle et al. (2015); Fasehun (2015); Mulamula and Amadi-Echendu (2017); Khan et al. (2017)
	Limited industrial capacity	Kennedy, Basu (2013); Karakosta and Psarras (2009); Flamos *et al.* (2010); Flamos (2010); Parnphumeesup and Kerr (2011); Mulamula and Amadi-Echendu (2017)
	Conflicting technical language	Lin, Berg (2001); Nasri *et al.* (2010); Pearce *et al.* (2012); Vac and Fitiu (2017); Manyuchi (2017); Eitzel *et al.* (2018); Chege *et al.* (2019)
	Risk aversion	Pérez, Sánchez (2003)
	Slow diffusion of innovative technology in markets	Worrell et al. (2001)
	Lack of technical knowledge	Koefoed, Buckley (2008)
Regulatory or policy barriers	High cost of investment in R&D / obtaining the technology	Karakosta and Psarras (2009); Ravindranath and Balachandra (2009); Karakosta, Doukas and Psarras (2010a); Jason *et al.* (2010); Parnphumeesup and Kerr (2011); Karakosta, Doukas and Psarras (2011); Doukas *et al.* (2012); Kruckenberg (2015b); Mulamula and Amadi-Echendu (2017); Khan, Haleem and Husain (2017)
	Lack or excess of regulation / Unclear legislation	Flamos *et al.* (2010); Flamos (2010); Karakosta et al. (2011); Azam (2011); Parnphumeesup, Kerr (2011); Doukas et al. (2012); Escalante et al. (2013); Belman and Tzachor (2013); Silva Jr. *et al.* (2013); Blohmke (2014); Fasehun (2015); Urban *et al.* (2015); Mulamula and

		Amadi-Echendu (2017)
	Lack of financial resources	Kennedy, Basu (2013); Koefoed, Buckley (2008); Flamos *et al. (2010);* Karakosta et al. (2010a); Flamos (2010); Jason et al. (2010); Doukas et al. (2012); Silva Jr. *et al.* (2013)
	Conflict of laws	Koefoed, Buckley (2008); Van Hoorebeek (2004)
	Differences in requirements between countries	Lin, Berg (2001); Karakosta et al. (2010a); Dasgupta and Taneja (2011); Doukas et *al. (2012)*; Karakosta et al. (2012)
	Insufficient resources devoted to technology transfer from universities	Siegel et al. (2003); O'Brien et al. (2007); Escalante et al. (2013); Belmonte et al. (2015); Urban *et al.* (2015); Khan et al. (2017)
	Political instability	Jason *et al.* (2010); Kennedy, Basu (2013); Kruckenberg (2015b); Chege *et al.* (2019)
	High capital costs	Kennedy, Basu (2013); Ockwell et al. (2008); Liu, Liang (2011)
	Financial capacities	Kennedy, Basu (2013); Karakosta et al. (2010a)
	Bureaucracy and inflexibility of universities	Kennedy, Basu (2013); Siegel et al. (2003)
	Inadequate tariff incentives	Kennedy, Basu (2013)
	High cost of regulatory compliance	Kennedy, Basu (2013)
	Limited access to capital	Kathuria (2002)
	Patents take too long	Van Hoorebeek (2004)
Personal or institutional barriers	Lack of competence and adequate training	Jason *et al.* (2010); Karakosta et al. (2010a); Flamos (2010); Nhamo (2011); Doukas et al. (2012); Escalante et al. (2013); Blohmke (2014); Adenle, Azadi and Arbiol (2015); Urban et *al.* (2015); Mehta *et al.* (2016); Mulamula and Amadi-Echendu (2017); Khan et al. (2017); Morgera and Ntona (2018)
	Lack of knowledge of new technology	Pérez, Sánchez (2003); Karakosta and Psarras (2009); Jason *et al.* (2010); Flamos (2010); Nhamo (2011); Doukas et al. (2012); Blohmke (2014); Adenle et al. (2015); Urban *et al.* (2015); Mehta et al. (2016); Mulamula and Amadi-Echendu (2017); Khan et al. (2017); Morgera and Ntona (2018)
	Lack of confidence	Pérez, Sánchez (2003); Karakosta and Psarras (2009); Van der Gaast *et al.* (2009); Karakosta, Doukas and Psarras (2009); Ravindranath and Balachandra (2009); Karakosta et al. (2010a); Flamos et al. (2010); Pearce *et al.* (2012); Doukas et al. (2012); Khan et al. (2017)
	Resistance to change	Koefoed, Buckley (2008); Karakosta, Psarras (2009); Ravindranath and Balachandra (2009); Silva Jr. *et al.* (2013); Khan et al. (2017); Chege *et al.* (2019)
	Difference in culture between companies	Lin, Berg (2001); Koefoed, Buckley (2008); Escalante et al. (2013); Belmonte *et al.* (2015); Mulamula and Amadi-Echendu (2017)
	Lack of information	Kennedy, Basu (2013); Worrell et al. (2001); Kathuria (2002); Koefoed, Buckley (2008); Ockwell et al. (2008)
	Inadequate information and	Kennedy, Basu (2013); Blohmke (2014); Urban *et al.*

	feedback	(2015); Mehta *et al.* (2016); Lema and Lema (2016)
	Geographic and cultural distance	Lin, Berg (2001); Flamos *et al.* (2010); Karakosta et al. (2011); Pearce *et al.* (2012)
	Shortage of professional institutions	Kennedy, Basu (2013); Blohmke (2014); Urban *et al.* (2015); Mulamula and Amadi- Echendu (2017)
	Problematic stakeholder interactions	Van Hoorebeek (2004)

Source: Own Authors (2021)

According to Greiner and Franza (2003), technical barriers are those that are present where there is a new technology, but for lack of knowledge it has never been applied before. Regulatory or political barriers involve laws and governmental procedures. And finally, the personal barriers are those that occur when the people involved have no technical knowledge towards the technology and no will to acquire this knowledge. According to the author, the latter is the most difficult to overcome.

One way that these barriers can be reduced and/or eliminated is the use of technology transfer mechanisms. When used in the most correct and effective manner possible, many of these barriers presented so far can be reduced.

4 MATURITY MODELS

To measure the level of development of an industry or some of its processes, maturity models are generally used as a support tool (SCHUMACHER et al., 2016) that describes the steps to improve a certain process, incorporating the components of each level and recommendations (PAULK, 2002).

A maturity model (MM) is a structured set of elements that describes an evolutionary path of improvement from immature processes to mature, effective and qualitatively better processes. Maturity can be captured qualitatively or quantitatively in a discrete or continuous manner (KOHLEGGER et al., 2009). Thus, maturity levels help in assessing and understanding the current state of the organisation, providing guidelines for continuous improvement within the organisation (KOSIERADZKA, 2017).

The main elements of a maturity model are (ISO, 2015; Lasrado, 2018):

(1) Dimensions: specific capability areas, process areas or design objects that structure the field of interest. Each dimension is further specified by a series of measures (practices, objects or activities) or by qualitative descriptions for each maturity level.

(2) Maturity levels: represent maturity states of a given dimension or domain. The characteristics of each level should be distinct and empirically testable, and the relationship of each level with its predecessor and successor should be well defined.

(3) Assessment tools: these can be qualitative or quantitative, for example using Likert-based questionnaires and scoring models.

(4) Boundary conditions: Particular conditions that organisations need to fulfil in order to progress from one level to another, are considered as the essential condition of a given maturity level.

(5) Stage boundaries: specific point at which the organisation advances to the next level.

(6) Path to maturity: a linear and progressive progression in which organisations develop and improve their capabilities, value creation, performance, etc. by walking along the path to maturity.

Although the terms *roadmaps*, maturity models, *frameworks* and/or readiness assessments may sound similar, they have a slight difference in their definitions:

- *Roadmaps* are "plans that correspond to short and long term objectives with specific technological solutions to help achieve these objectives" (GARCIA; BRAY,

1997).

- Maturity models are models that help an individual or entity to reach a more sophisticated maturity level (i.e. skill) in people / culture, processes / structures and / or objects / technologies, following a step-by-step process of continuous improvement (METTLER, 2011).
- *Frameworks* are collections of coherent procedures, methods and tools to architect (i.e. design and engineer) a system (STOREY, 2005).
- Readiness assessments are evaluation tools to analyse and determine the level of readiness of conditions, attitudes and resources, at all levels of a system, required to achieve its objective (s) (BENEDICT et al., 2017).

4.1 I4.0 MATURITY MODELS

In the existing literature, there are still few proposals regarding maturity models for the implementation of Industry 4.0. These models are a guide for the implementation of Industry 4.0 in order to mitigate misinterpretations.

In the literature review carried out, 34 articles were found that presented maturity models in I4.0. However, 10 of them proved to be more relevant to the research, fitting the proposal better and presenting a higher InOrdinato (See section 6.1) than the others.

Next, summaries of each of these 10 selected models are presented. After discussing them individually, in Section 5.3 a comparative table is presented, with the objective of obtaining a clearer view of the behaviour of each model (Table 9).

MM1. A Categorical Framework of Manufacturing for Industry 4.0 and Beyond (Qin; Liu, 2016).

In Qin and Liu's (2016) model, the author states that the production system can be divided into three levels of automation: machine, production process, and factory system. By combining the intelligence level with the engineering level, a hierarchical structure (Figure 9) is generated with a total of nine intelligence applications, with three intelligence level technologies acting in the three engineering level sections. These nine applications range from low intelligence and simple automation to high intelligence and complex automation.

From applications I to IX, the production system becomes more and more automated, flexible and intelligent. It is necessary to know that the high-level performance targets and technologies are based on the low level, which means that this structure works in sequence. The IX application is considered as the implementation of industry 4.0.

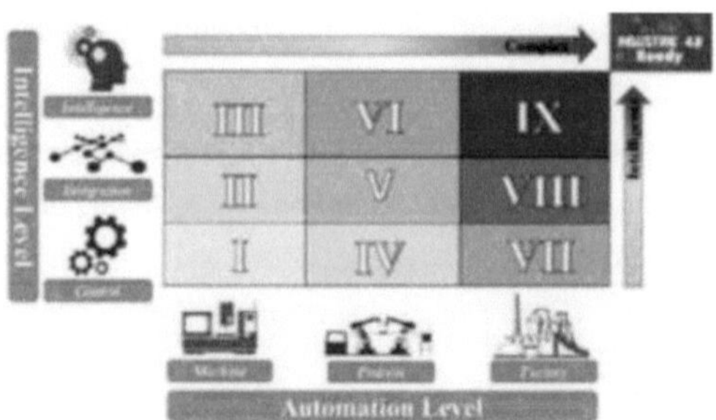

Figure 9 - Qin and Liu's (2016) Maturity Model
Source: Qin and Liu (2016)

MM2. A maturity model for assessing Industry 4.0 readiness and maturity of manufacturing enterprises (Schumacher et al., 2016)

The Industry 4.0 Maturity Model by Schumacher et al. (2016) aims at an extension of existing models and tools through its strong focus on organisational aspects. The authors sought to transform the abstract concepts of smart manufacturing into items that can be measured in real production environments.

To facilitate different analyses of Industry 4.0 maturity, the proposed model includes a total of 62 maturity items, grouped into nine enterprise dimensions. Table 5 provides an overview of the dimensions, along with some exemplary items for better understanding.

Table. 5 - Dimensions and maturity items of the model by Schumacher et al. (2016)

Dimension	Example of a maturity item
Strategy	Implementation of Roadmap I4.0, Resources available for realization, Adaptation of business model...
Leadership	Will of leaders, Management of competencies and methods, Existence of central coordination for I4.0,...
Customers	Use of customer data, Digitisation of sales/services, Competence in customer digital media,...
Products	Product individualisation, product digitalisation, interaction of products with other systems,...
Operations	Decentralisation of the process, Modelling and simulation, Interdisciplinarity,...
Culture	Knowledge transfer, Open innovation, Value of ICT in the firm,...
People	ICT skills of employees, Openness of employees to new technologies, Autonomy of employees,...

Governance	Labour regulations for I4.0, Adequacy of technological standards, Protection of intellectual property, ...
Technology	Existence of modern ICT, Use of mobile devices, Use of machine-to-machine communication, ...

Source: Schumacher et al. (2016)

The evolution path for each item goes through five maturity levels, where level 1 describes a complete lack of attributes that support Industry 4.0 concepts and level 5 represents the state of the art of the required attributes.

The maturity assessment within a company is conducted using a standardised questionnaire consisting of one closed question per item. Each question requires a response to a Likert-type scale ranging from 1 "not distinct" to 5 "very distinct".

MM3. Industrie 4.0 Maturity Index - Managing the Digital Transformation of the Companies.(SCHUH et al. 2017)

Schuh et al. (2017) considered "digitisation" as an enabler for Industry 4.0 and developed a maturity index with the help of a four-stage methodology.

The maturity index considered six stages of development which are: (1) Computerisation, meaning the installation of computers (and information systems) that will support tasks with data processing systems and relieve employees of repetitive manual activities; (2) Connectivity, referring to the connection of these computers and data processing systems to support key business processes; (3) Visibility, creating a digital shadow of what is happening in real time in the factory and supporting management decision with data; (4) Transparency, enabling visualisation of events and understanding the root cause of their happening; (5) Predictability, building on the understanding of the Transparency stage and moving forward to plan and make decisions based on future scenarios; and (6) Adaptability, including the autonomous response of machines and other systems based on their predictive capability.

It is important to highlight that the stages of computerisation and connectivity are more related to the objectives of the industry 3.0 era, while the industry 4.0 journey starts at the visibility maturity stage. Furthermore, four key areas, each with two subdimensions, were assessed at each maturity stage: (a) resources (digital capability and structured communication), (b) information systems (self-learning information processing and information system integration), (c) organisational structure (organic internal organisation and dynamic collaboration within the value network); and (d) organisational culture (willingness to change and social collaboration).

The aim of the paper was to define a maturity index, which can assess the current stage of Industry 4.0 of the organisation and find the measures, which can lead them to a higher maturity stage. The idea of a customized view of Industry 4.0 for a company was also emphasized.

MM4. Three Stage Maturity Model in SME's towards Industry 4.0 (Ganzarain; Errasti, 2016)

In Ganzarain and Errasti's Model (2016), the following maturity scale was defined for the three-stage process model: (1) Initial: there is no specific vision of the company's industry 4.0; (2) Managed: there is a roadmap of the industry 4.0 strategy; (3) Defined: customer segments, value proposition and key resources defined; (4) Transformation: Transforms the strategy into concrete projects; (5) Detailed BM: Transformation of the Business Model.(Figure 10)

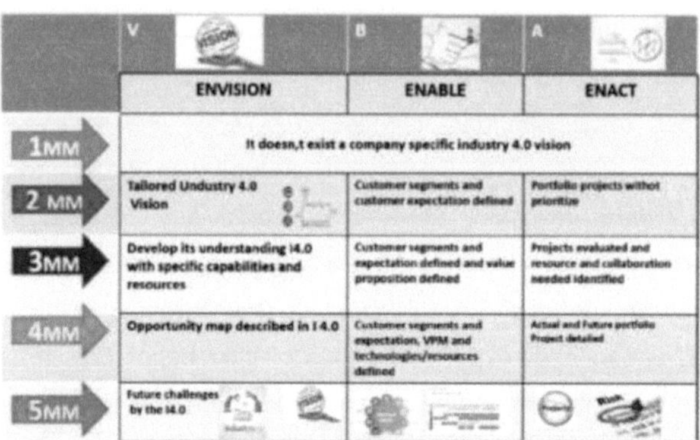

	V ENVISION	B ENABLE	A ENACT
1MM	It doesn,t exist a company specific industry 4.0 vision		
2 MM	Tailored Undustry 4.0 Vision	Customer segments and customer expectation defined	Portfolio projects withot prioritize
3MM	Develop its understanding I4.0 with specific capabilities and resources	Customer segments and expectation defined and value proposition defined	Projects evaluated and resource and collaboration needed identified
4MM	Opportunity map described in I 4.0	Customer segments and expectation, VPM and technologies/resources defined	Actual and Future portfolio Project detailed
5MM	Future challenges by the I4.0		

Figure 10 - Ganzarain and Errasti (2016) Maturity Model
Source: Ganzarain and Errasti (2016)

The *Vision 4.0* stage is dedicated to defining a customised vision of Industry 4.0, developing your own understanding of the general ideas of I4.0 with company-specific resources and capabilities. In this phase, external experts and technology partners are involved to show relevant results and best practices, and to support the process of building the Industry 4.0 vision. The result of this stage is a personalised vision of the company for a future end state, supporting the future challenges proclaimed by the new industrial revolution.

In the *Enable* stage, the company starts from a substantial vision of Industry 4.0

and, based on this vision, tries to define the technology portfolio and the resources needed to support the new product service solutions identified in the previous stage. The 4.0 *Roadmap* will facilitate the strategy planning and alignment process to visualise and structure the different strategies. There are four strategic perspectives: market, product, process and value network. The result of this stage (*Enact*) is a timely and multi-pronged map of the overall strategy towards the Industry 4.0 vision.

MM5. Development of an Assessment Model for Industry 4.0: Industry 4.0-MM (Gökalp et al. 2017)

The Industry4.0-MM framework is formed on the basis of ISO / IEC 15504 - also referred to as software process improvement and capability elimination (SPICE). The aim of the authors was to create a common basis for evaluating the establishment of Industry 4.0 technologies and to present the evaluation results using a common rating scale. However, instead of the process dimension of SPICE, the Aspect dimension was assigned, which can be seen in Figure 11.

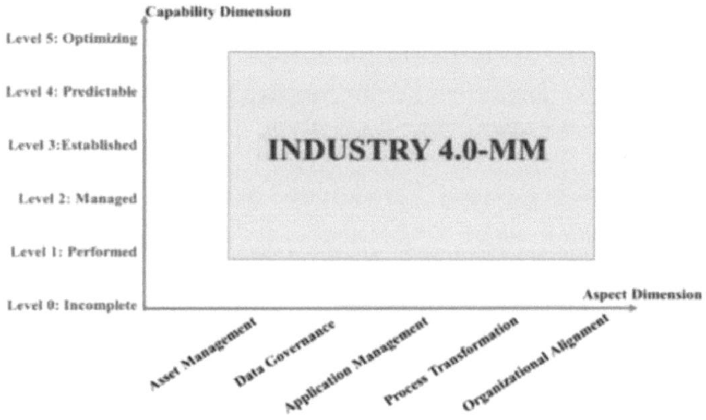

Figure 11 - Maturity Model by Gokalp et al. (2017)
Source: *Gökalp et al. 2017*

It is difficult to identify the limits of the transformation process in Industry 4.0. Besides process transformation, other dimensions such as infrastructure, information systems, data and organisation, as well as their integration, are critical in the establishment of Industry 4.0. Consequently, a new arrangement of Industry 4.0 processes and practices is performed to integrate them under meaningful and compatible abstract definitions, known as "Aspects".

In the aspect dimension, aspects are defined and classified into categories such as Asset Management, Data Governance, Application Management, Process Transformation and Organizational Alignment. The capacity dimension is defined by capacity levels and capacity indicators. The capability dimension is adopted from SPICE, it has 6 levels, from "Level 0: Incomplete" to "Level 5: Optimizing".

MM6. Smart Factory Implementation and Process Innovation (Sjodin et al., 2018).

Sjodin et al. (2018) revealed three general principles underlying a successful I4.0 implementation: *Cultivate "digital people"; Introduce agile processes; Configure modular technology.*

Within these three areas - people, process and technology - interview data identified key activities that underpin the development of smart factory capabilities; this framework fits the dimensions defined by previous studies on change management. The authors categorised these key activities by maturity level to create a smart factory maturity model (Table 7). Figure 12 presents this model proposed by the authors.

Table. 6 - Maturity Model by Sjodin et al. (2018)

Maturity Level	People Cultivating digital people	Processes Introduce agile processes	Technology Configuring modular technology
Level 4. Predictable Smart Manufacturing	Create a culture of continuous smart factory innovation. Create specialised roles and responsibilities focused on predictable production.	Develop processes to integrate data visualisation into decision making. Create proactive processes to forecast and plan future production.	Create systems to monitor and visualise critical operational analysis. Integrate digital system information from external partners to enable supply chain predictability.
Level 3. Real-time process analysis and optimisation	Organise sense-making sessions with suppliers, users and other stakeholders. Recruit data analysts and data scientists to optimize production.	Use insight analysis and data interpretation to optimise operational processes. Create processes to assess opportunities for optimisation.	Implement systems for real-time performance analysis. Implement simulation systems to test, prototype and optimise the digital factory.
Level 2. Structured data collection and sharing	Educate people to develop the capacity to exploit connected data systems. Review production team roles to proactively coordinate digital insights and knowledge sharing.	Create specialised insight mining processes to support cross-departmental information gathering. Create cross-functional scanning networks to facilitate knowledge sharing.	Increase the accuracy of technology data collection. Create automated processes for data mining and sharing across functions.
Level 1. Connected technologies	Create an inclusive culture for implementation by involving the workforce in developing the vision. Recruit people with digitalisation skills.	Formalise the processes for implementing the hybrid smart factory. Create process to involve external stakeholders in the development of the connected platform.	Apply a digital lens to map new and existing technologies. Connect existing technology applications to create data flow.

Source: Sjodin et al. (2018)

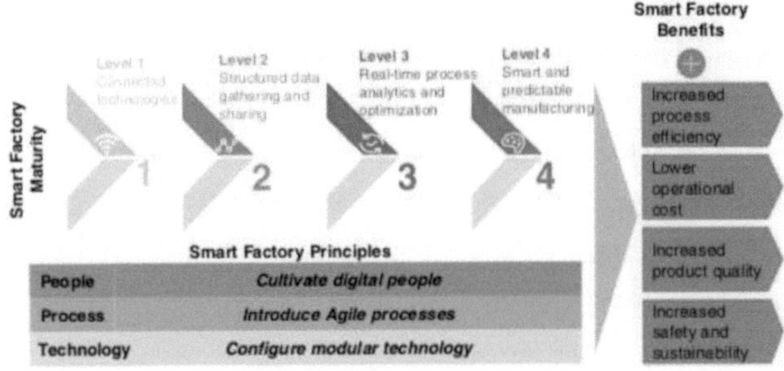

Figure 12 - Maturity Model by Sjodin et al. (2018)
***Source:* Sjodin et al. (2018)**

MM7. Maturity and Readiness Model for Industry 4.0 Strategy (AKDIL et al. 2018)

Akdil et al. (2018) proposed an Industry 4.0 maturity model with four maturity levels and three broad dimensions. The levels considered are absence (level 0 - Industry 4.0 requirements are not met), existence (level 1 - the company's use of integration, automation, data collection, digital technologies, interoperability etc. is at a very low level), survived (level 2 - the company's use of integration, data sharing, interoperability etc. are at a medium level) and maturity (level 3 - the company's use of integration, data sharing, interoperability etc. is at a high level).

Furthermore, Akdil et al. (2018) recognised various principles and technologies for Industry 4.0. Thus, three dimensions (smart products and services, smart business processes and strategy and organisation) and their corresponding maturity levels (level 0-3) were defined in this model with the help of a set of key characteristics. Finally, a survey-based instrument was used to find the maturity level and a demonstration of the retail sector was presented.

MM8. Roadmapping towards industrial digitisation based on an Industry 4.0 maturity model for manufacturing enterprises - A new methodology to analyse the functional and physical architecture of maturity model for manufacturing enterprises (Schumacher et al., 2019)

For the development of the Schumacher et al. (2019) maturity model, the following three main requirements were derived, which the new model should meet:

- Operationalisation of the abstract concepts of Industry 4.0 (such as cyber-physical system, embedded systems or vertical / horizontal integration), as professionals have problems to apply and evaluate these concepts in their company;

▪ Definition and description of each maturity level for each maturity item separately, such as very extensive ratings among respondents when using generic scales (e.g. Likert scales) without clear reference;

▪ Transformation of the maturity results into a maturity report, including the authors / experts' interpretation of the assessment results, as companies have problems deriving the next steps from independent charts and diagrams due to the complexity and novelty of the topics.

The following development approach was followed, whereby phases 1-3 focus on developing the maturity model and phases 4-6 on developing the final realisation model (Figure 13).

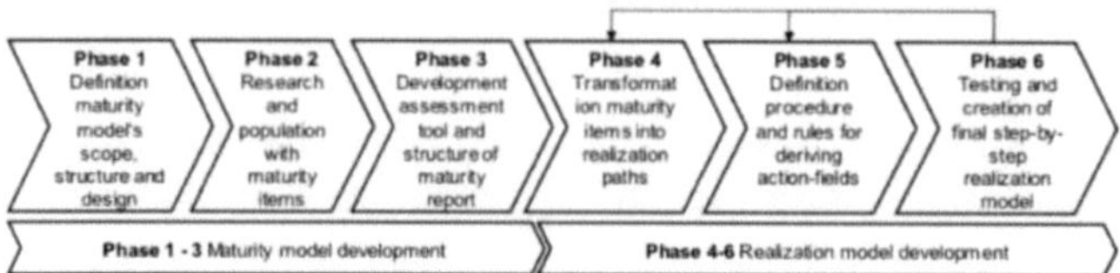

Figure 13 - Maturity Model by Schumacher et al. (2019)
Source: Schumacher et al. (2019)

The third requirement for the new model marks the starting point for development phases 4-6, in which the maturity items are related and logically sequenced to each other to create realisation paths for Industry 4.0. Therefore, the maturity items were assigned to three groups related to three realisation phases in Industry 4.0:

▪ Enable: contains maturity items that build the foundations for the realisation of Industry 4.0, such as sensors on machines or the willingness of employees to use modern ICTs.

▪ Implement: contains maturity items that capture the endorsement of Industry 4.0 concepts, such as analysing the data collected or conducting employee training.

▪ Formalise: contains maturity items that help sustain target states in Industry 4.0, such as defined technology standards or digital information processes within the company.

By inserting logical relationships between the 65 maturity items, depending on the necessary chronological order in which companies should direct these items, a generic achievement roadmap for Industry 4.0 was created (Figure 12 - Phase 4). In phase 5, a step-by-step procedure was defined and a set of rules for using the results of the company's maturity assessment for its development. Finally, in phase 6, all resulting methods, tools and necessary steps are formalised in the Industry 4.0 realisation model.

MM9. A Maturity Model for Logistics 4.0: An Empirical Analysis and a Roadmap for Future Research (Facchini et al. 2020)

The proposed model is designed to offer measures that can be translated into a set of guidelines or recommended solutions for Logistics 4.0. It emphasises the need to manage the entire set of material and information flows. In this sense, the authors identified three main dimensions, characterised by dedicated assessment areas:

(1) Management: Investments, innovation management, value chain integration.

(2) Material flow: Degree of automation and robotisation in warehouse and transportation, Internet of things, 3D printing, 3D scanning, advanced materials, augmented reality, smart products.

(3) Information flow: Data-based services, Big data (data capture and use), RFID, RTLS (real-time location systems), IT systems (ERP, WMS, cloud systems).

These three dimensions can be used to assess the maturity and awareness of Industry 4.0 solutions for a Logistics 4.0 transition. The maturity levels are five: the first level identifies the absence of any Logistics 4.0 capability and the fifth level identifies the full implementation and integration of Logistics 4.0 solutions.

MM10. Development of maturity model for assessing the implementation of Industry 4.0: learning from theory and practice (Wagire et al., 2020)

Wagire et al. (2020) conducted an Industry 4.0 maturity model that is empirically grounded and technology-centric to assess the maturity level of Indian manufacturing organisations. The model consists of 7 dimensions and 38 maturity items. 'People and culture', 'industry 4.0 awareness', 'organizational strategy', 'process and process chain', 'smart manufacturing technology', 'product and service oriented technology' and 'industry 4.0 core technology' are the dimensions comprised in the maturity model to assess the maturity level of the organization. The maturity items and dimensions are prioritized based on their importance levels resolved using the Fuzzy Analytic Hierarchy Process (FAHP). The model is tested, validated and implemented in five Indian manufacturing companies. Figure 14 presents the model developed by the authors, and how the result of one of the companies in which it was applied looked like.

Figure 14 - Wagire et al. (2020) Maturity Model
***Source:* Wagire et al. (2020)**

4.2 TT MATURITY MODELS

In terms of Technology and Knowledge Transfer, the existing maturity models are even more scarce when compared to I4.0.

In the literature review carried out, 16 articles were found in the search for TT maturity models. However, only 7 of them actually present a maturity model consistent with the present study. These 7 articles were classified according to their InOrdinatus (See section 6.1), being MM1 the one that presents a higher index, and MM7 the one that presents the lowest.

Next, summaries of each of these 7 selected models are presented. After discussing them individually, in Section 5.3 a comparative table is presented, with the objective of obtaining a clearer view of the behaviour of each model (Table 10).

MM1. A practice-based maturity model for hostile TTO performance management: development and initial use (Kreiling and Bounfour, 2019)

This paper presents the development and initial use of a practice-based maturity model for technology transfer organisations (TTOs). The intention is not to compare TTOs, but to find out whether there is a link between the maturity of TTO practices and organisational resources, competencies and context, as well as output and outcomes. Based on a conceptual framework for holistic TTO performance measurement, the model was refined and validated with TTO managers. It offers a new way for TTOs to determine the maturity of their practices in six areas: 'opportunity awareness and exploitation', 'boundary spanning', 'translation and combination', 'co-creation and development', 'cultural change management' and 'knowledge management'. These areas correspond to six TTO resources that are assessed with 44 practice statements.

The model was developed in six steps. Steps one and two (identification of appropriate texts and data sources, reading and categorisation) correspond to the literature review conducted. The identification of concepts and their categorization was the object of stages three and four. Stage five was the definition of maturity levels, and stage six the validation of the model.

MM2. Knowledge sharing maturity model for Jordanian construction sector (Arif et al., 2017)

The model of Arif et al. (2017) focuses on the analysis of the maturity of Knowledge Transfer in the construction sector. For the development of the model, the relationships and interactions between the cultural variables were considered. The impact of the variables was incorporated into the three levels of maturity. The first level identifies that the variable almost does not exist in the company's knowledge transfer practices. The second level shows the occasional techniques that the company uses to increase knowledge transfer activities. The final level demonstrates the importance of the variable in affecting knowledge transfer as being fundamentally rooted in the company's vision, mission, strategy and operations. The variables are presented in Table 7.

Table. 7 - Variables of the Maturity Model of Arif et al. (2017)

	Motivation: How does the company motivate employees to share knowledge?

	Management commitment: how do senior managers support knowledge sharing practices to provide a suitable environment for KS practices in the workplace?
	Leadership behaviour: How do leaders behave to encourage and support employees to share knowledge?
	Power distance: The extent to which members of a society accept that power in institutions and organisations is and should be distributed unequally.
	Uncertainty avoidance: the degree to which members of a society are uncomfortable with uncertainty and ambiguity and supports beliefs that promise dimensions that will affect the sharing of new information among members of the enterprise.
	Organisational structure: division of tasks among individual employees, groups or departments and locations. To control the work of an entity, methods and procedural measures are adopted that support KS activities.
	After refinement: Organisational form (family business): how do family members in the business affect knowledge sharing activities and how do employees react with them in terms of knowledge exchange?
Communication Variables	**Automonia:** The extent to which an individual or group of individuals has the freedom, independence and discretion to determine what actions are required and how best to carry them out, and how this kind of freedom affects KS activities in the company.
	Before refinement: employee relationships (outside the company): the social activities that employees undertake outside the company to strengthen the connection between them to increase KS and the company function supports these activities.
	Communication technology: the amount of communication technology the company provides to increase KS between employees, such as laptop, phone, fax, PC and internet.
	Networking: the interaction between groups of people who share a common interest; using social contacts to network. Using Internet networking groups to connect and communicate with each other for faster and easier access to information exchange.
	Gender differences: focuses on the degree to which society reinforces or fails to reinforce the traditional male work model of male conquest, control and power, which affects women in knowledge sharing.
	Collective Achievements: Focuses on the degree to which society reinforces collective achievements and interpersonal relationships.
Confidence Variables	**Mutual trust between employees**
	After refinement: employee relations (outside the company): the social activities that employees undertake outside the company to strengthen the connection between them to increase KS and the role of the company to support these activities.
	Before refinement: Organisational form (family business): how do family members in the business affect knowledge sharing activities and how do employees react with them in terms of knowledge exchange?

Source: Arif et al. (2017)

MM3. A mechanism for sharing best practices between university technology transfer offices (De Beer et al., 2017)

The model by De Beer et al. (2017) is called the *Improved Maturity* Model (IMM), and is an enhancement of the maturity model proposed by Secundo et al. (2016). This model aims to formalize a mechanism through which best practices can be identified and shared more effectively among TTOs.

The Enhanced Maturity Model (IMM) is presented to reflect the true performance and characteristics of the TTO. Each maturity level provides a description of the characteristics associated with that level that can be used by the TTO. Thus, strategic decisions can be made on how to improve certain areas and underpin the decision-making process more effectively. After self-assessing each efficiency area, a TTO can then focus on specific intangible indicators that show weaknesses in its efficiency and can reach a higher maturity. The defined, managed and integrated stages are divided into early and late stages to highlight the different characteristics of TTOs at these transition levels.

MM4. Measuring university technology transfer efficiency: a maturity level approach (Secundo et al., 2016)

The Maturity Model is inspired by Berkley's Model (PM) 2, which allows an organization to determine strengths and weaknesses and focus on weak practices to achieve greater maturity. The fuzzy analytical hierarchy process is adopted to determine the priorities and weights of non-monetary indicators as they are ambiguous.

The Maturity Model for measuring TTO effectiveness covers the following areas of effectiveness: intellectual property strategy and policy; organisation design and structure; human resources; technology; industry links; and *networking*. The model provides a theoretical context along which the maturity process can be incrementally developed in the TTO from one level to the next, moving from awareness, definition, management, integration and sustainment.

Each maturity level, provides a description of the characteristics associated with that level that can be used by the TTO to make strategic decisions on how to improve certain areas and underpin the decision-making process more effectively.

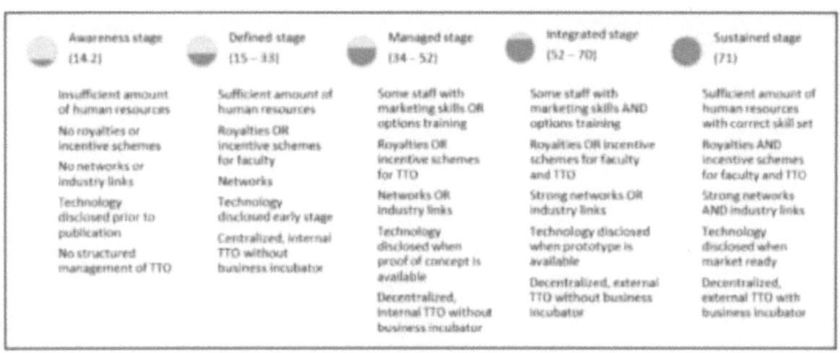

Figure 15 - Maturity Model by Secundo et al. (2016)
Source: Secundo et al. (2016)

MM5. Maturity measurement of knowledge-intensive business processes (Sinha et al., 2011)

Sinha et al.'s (2011) maturity model focuses on analysing knowledge across companies, and for SMEs to measure and assess the quality of their business processes.

The proposed maturity model categorises its indicators into seven key process areas (KPA), most of which are derived from the EFQM model. These are:

leadership, policy and strategies, partnership and resources, process design, knowledge transfer and design, employees, information system and two specific process areas. Each KPA is assigned several success factors consisting of knowledge, process and quality-oriented indicators. The process-specific KPA has its own success factors and can be expanded at will.

The maturity model is derived from existing established approaches and distinguishes five different levels: (1) Initial: knowledge intensive process with a non-formal / spontaneous character regarding process design and knowledge handling; (2) Repeated: proactive knowledge intensive process with a non-formal / personnel related character regarding process design and knowledge handling; (3) Defined: Established knowledge intensive process with a formal character; (4) Managed: controlled knowledge intensive process with formalised and proven character; (5) Optimised: sustainable knowledge intensive process; optimised and quality oriented process design with continuous improvement.

The result of the maturity level determination allows SMEs to derive improvement actions for skill development in relation to knowledge handling and quality oriented process design.

The following iterative phases show the practice of the proposed maturity model:

- Self-assessment. The participants in the process analyse a set of defined instructions and set out their estimates appropriately.
- Identification of measures for improvement. During the discussion in the course of the analysis of the possible improvement of the declaration catalogue, the responsibilities and corresponding deadlines of the measures are documented.
- Implementation of identified improvement actions. The identified improvements are implemented and appropriate changes are made in the process. After this phase, a periodic self-assessment (Phase 1) should be carried out to ensure that the process is in line with the concept of continuous process improvement.

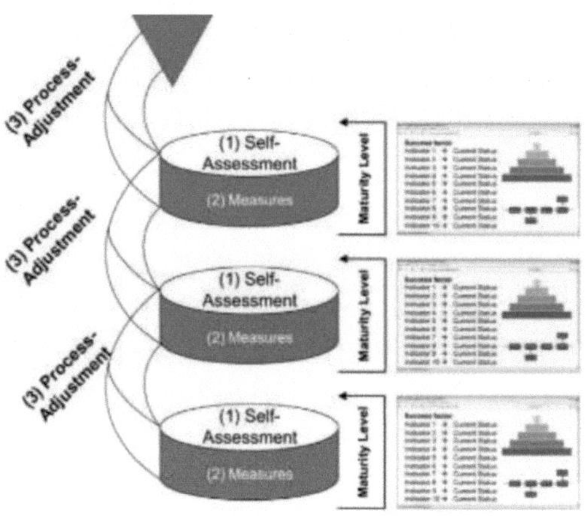

Figure 16 - Sinha et al. (2011) Maturity Model
Source: Sinha et al. (2011)

MM6. Roadmap for knowledge sharing and transfer: sustaining outsourcing relationships (Phusavat; Kess, 2008)

Phusavat and Kess's (2008) model aims to develop a roadmap designed to assist in the process of planning knowledge transfer activities.

In the model there are a total of three main stages, based on the treatment of knowledge sharing and transfer to suppliers. They are: (1) preparation and communication with teams, (2) organisation, and (3) integration.

In the early stage, the focus is on preparing the company in terms of specific work, activities and tasks to be undertaken to initiate knowledge sharing and transfer. From his perspective, it is essential that constant updates and communications with teams are made clear. The next step is to organise these jobs, activities and tasks. Finally, it is crucial that activities / practices / tasks across knowledge layers are integrated into a management process. They should be routinely and regularly performed simultaneously. Feedback and progress are explicitly contained in management reports.

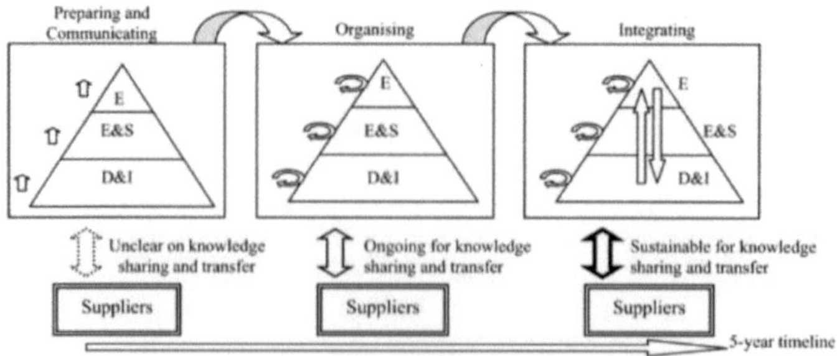

Figure 17 - Phusavat and Kess (2008) Maturity Model
Source: Phusavat and Kess (2008)

MM7. Knowledge and Technology Transfer: Levels and Key Factors (Sung and Gibson, 2000)

Sung and Gibson (2000) developed a model where four levels of Knowledge and Technology Transfer are presented: Creation (Level I), Sharing (Level II), Implementation (Level III) and Commercialisation (Level IV).

Figure 18 - KTT levels from Sung and Gibson (2000)
Source: Sung and Gibson (2000)

The comprehensive literature identifies sixteen variables that affect the process and outcomes of knowledge and technology transfer. Research results show four key factors in knowledge and technology transfer: Communication, Distance, Equivocality and Motivation.

Communication refers to the degree to which a medium is able to efficiently and accurately transmit task-relevant information and media, whereas distance involves physical and cultural proximity. Equivocality refers to the degree of concreteness of the knowledge and technology to be transferred, while motivation involves incentives and recognition of the importance of knowledge and technology transfer activities.

4.3 ANALYSIS OF THE MATURITY MODELS

After the models have been studied individually, a comparative table was made with all of them. In this table it is possible to analyse the main context of each model, the maturity levels and dimensions present in each of them. Chart 8 presents the I4.0 maturity models.

Box. 8 - I4.0 Maturity Models

Model	Author	Background to the research	Maturity Levels	Dimensions
MM1	Qin et al. (2016)	Framework of automation levels for I4.0	9 maturity levels, analysing intelligence and complexity.	6 dimensions: (1) Intelligence, (2) Integration, (3) Control, (4) Machine, (5) Production Process, (6) Factory System
MM2	Schumacher et al. (2016)	I4.0 Maturity Model for manufacturing companies	5 maturity levels: (1) complete lack of attributes that support 4.0 concepts; (5) state of the art of the necessary attributes	9 dimensions: (1) strategy, (2) leadership, (3) customers, (4) products, (5) operations, (6) culture, (7) people (employees), (8) governance and (9) technology.
MM3	Schuh et al. (2017)	I4.0 Maturity Model for manufacturing companies	6 maturity levels: (1) computerisation, (2) connectivity, (3) visibility, (4) transparency, (5) predictive capacity, (6) adaptability.	4 dimensions: (1) resources, (2) information systems, (3) organisational structure and (4) organisational culture.
MM4	Ganzarain; Errasti (2016)	I4.0 Maturity Model for SMEs	5 maturity levels: (1) Initial, (2) Managed, (3) Defined, (4) Transformation, (5) Detailed BM.	The dimensions are not shown in the article.
MM5	Gokalp et al. (2017)	I4.0 Maturity Model for manufacturing companies	6 maturity levels: (0) Incomplete, (1) Accomplished, (2) Managed, (3) Established, (4) Predictable, (5) Optimizing.	5 dimensions: (1) Asset management, (2) Data management, (3) Application management, (4) Organizational alignment, (5) Process transformation.
MM6	Sjödin et al. (2018)	I4.0 Maturity Model for SMEs	4 maturity levels: (1) connected technologies, (2) structured data collection and sharing, (3) real-time process analysis, (4) predictable Smart Manufacturing	3 dimensions: (1) People, (2) Processes, (3) Technologies.

MM7	Akdil et al. (2018)	I4.0 Maturity Model for manufacturing companies	4 levels of maturity: (0) absence, (1) existence, (2) survived and (3) maturity.	3 dimensions: (1) smart products and services, (2) smart business processes and (3) strategy and organisation.
MM8	Schumacher et al. (2019)	Roadmap for I4.0 in companies	3 levels of maturity: (1) Activate, (2) Implement, (3) Formalise.	The dimensions are not shown in the article.
MM9	Facchini et al. (2020)	Maturity Model for Logistics 4.0	5 maturity levels: (1) ignoring, (2) defining, (3) adopting, (4) managing, (5) integrating	3 dimensions: (1) Management, (2) Material Flow, (3) Information Flow
MM10	Wagire et al. (2020)	I4.0 Maturity Model for manufacturing companies	5 levels of maturity: (1) layman, (2) digital novice, (3) experienced, (4) expert.	7 dimensions: (1) people and culture, (2) I4.0 awareness, (3) organisational strategy, (4) process chain, (5) smart manufacturing technology, (6) product and service oriented technology, (7) I4.0 core technology.

Source: Own authorship (2021)

The I4.0 maturity models analysed are quite different from each other. Each one of them follows a line of reasoning and aims at different objectives. Some focus on Small and Medium Enterprises, others on the maturity of the technology required in I4.0, and still others on the maturity of I4.0 as a whole. Only one of them presents a connection with Technology and Knowledge Transfer, being MM6 which presents a Knowledge Management dimension. But none of the articles presents a model where the focus is on I4.0 and TT. Table 9 presents the maturity models of Technology and Knowledge Transfer.

Box. 9 - KTT Maturity Models

Model	Author	Background to the research	Maturity Levels	Dimensions
MM1	Kreiling, Bounfour (2019)	Technology Transfer in TTOs	5 levels: (1) B-Initial, (2) B-Managed, (3) A-Defined, (4) AA-Generalised, (5) AAA-Advanced.	5 dimensions: (1) Resources and competencies, (2) Organizational parameters, (3) Generic practices, (4) Production, (5) Results
MM2	Arif et al. (2017)	Knowledge Transfer	3 levels: (1) barely exists, (2) occasional techniques that the company uses, (3) fundamental TC.	3 dimensions: (1) management variables, (2) communication variables, (3) trust variables.
MM3	Beer et al. (2017)	Technology Transfer in TTOs	8 levels: (1) awareness, (2) initially defined, (3) final defined, (4) initially managed, (5) final managed, (6) initially	6 dimensions: (1) Human Resources, (2) Strategy and Policies, (3) Networking, (4) University-Industry Links, (5) Technology, (6) Organisation Design and

			integrated, (7) final integrated, (8) sustained.	Structure.
MM4	Secundo et al. (2016)	Technology Transfer in TTOs	5 levels: (1) awareness, (2) defined, (3) managed, (4) integrated, (5) sustained.	6 dimensions: (1) Human Resources, (2) Strategy and Policies, (3) Networking, (4) University-Industry Links, (5) Technology, (6) Design and structure.
MM5	Jochem et al. (2011)	Knowledge between companies	5 levels: (1) initial, (2) repeated, (3) defined, (4) managed, (5) optimised.	7 dimensions: (1) leadership, (2) policy and strategies, (3) partnership and resources, (4) process design, (5) knowledge transfer and design, (6) employees, (7) information system.
MM6	Phusavat; Kess (2008)	Knowledge Transfer	3 levels: (1) preparation and communication, (2) organisation, (3) integration	2 dimensions: (1) explicit knowledge, (2) tacit knowledge
MM7	Sung, Gibson (2000)	Levels of Knowledge and Technology Transfer	4 levels: (1) Creation, (2) Sharing, (3) Implementation, (4) Commercialisation.	4 dimensions: (1) Communication, (2) Distance, (3) Equivocality, (4) Motivation

Source: Own authorship (2021)

One notices the small number of studies that refer to maturity models of Technology and Knowledge Transfer. Of these, three focus primarily on knowledge transfer and three on technology transfer at TTOs. One of them presents the Knowledge and Technology Transfer levels. Thus, one can conclude the similarities that exist between them and the strong gap in the literature: there is no model that can measure the maturity of companies in relation to all components that aggregate the Transfer of Technology and Knowledge.

5 TECHNOLOGY TRANSFER IN INDUSTRY 4.0

Technology Transfer can occur in several ways. An important distinction in the literature on TT is between vertical and horizontal technology transfer. Both are considered important for economic development, and can be present together in an organisation. In horizontal technology transfer, knowledge is diffused to competitors, while in vertical technology transfer, knowledge is diffused to potential input suppliers (OCKWELL et al., 2008).

Industry 4.0 is a new and important topic for organisations. Many companies are implementing these new technologies belonging to this new industrial revolution. One way to facilitate this implementation process of Industry 4.0 in organizations, is to use Technology Transfer activities and mechanisms (DA SILVA et al., 2019). This relationship can be in several ways, and it is fundamental to understand how it can happen.

Therefore, based on the literature review conducted, the different ways in which the relation between Industry 4.0 and Technology Transfer might occur, were classified. Eight different relations were found (R1-R8), presented in Figure 19.

Figure 19 - Relations between Industry 4.0 and Technology Transfer
Source: Own authorship (2021)

5.1 TEACHING FACTORIES

These transformations that are taking place in modern society generate an additional challenge for education. Future employees will need various skills to deal with this new reality of industry and society (MAIA et al., 2017). Industry 4.0 propagates the idea of workers who will increasingly focus on creative, innovative and communicative activities. In addition, work with complex problem solving, systemic thinking and social skills. In social skills is under consideration persuasion, emotional intelligence, teaching skills and active learning (LANZA et al., 2015).

The development of *Learning Factories* is a critical concept in the reindustrialisation of Europe. According to Bedolla et al. (2017), the most industrialised countries are funding national and international programmes to promote the integration of Industry 4.0 enabling technologies into production environments.

The Learning Factories offer a realistic environment of production systems by the use of their technical equipment. Furthermore, process improvements and modifications can be tested safely during production processes. Therefore, Learning Factories offer a great opportunity for training and preparing students for the use of Industry 4.0 (KRÜCKHANS et al., 2016) and should be closely linked to cyber-physical production systems (ABELE et al., 2015).

5.2 MAN-MACHINE INTERACTION

In addition to industry developments towards the digitisation of manufacturing, the technological momentum of Industry 4.0 forces organisations to adopt an adaptive behaviour, leading them to new management processes promoted by evolutions in learning. Industry 4.0 will affect the way businesses related to human resource management and human trends are considered (LIBONI et al., 2019).

Bahrin et al. (2016) identify the interconnectedness of humans, objects and systems, leading to optimised and self-organised real-time dynamic value systems as one of the critical components of Industry 4.0. Workers add a lot of value to innovation and organisational learning processes. The modern manufacturing system is considered a composition of cyber-physical, cybernetic and human components, and IoT is used as a connecting element for integration (THRAMBOULIDIS; CHRISTOULAKIS, 2016).

This interconnection of man with the objects and systems of I4.0, is called Human-Machial Interface (HMI). This relationship is characterised by the closeness of the relationship between the system and the human operators, who may share the same workspace. According to this vision, the integration of human agents is considered from the initial stage of the design of the artificial system and the entities that will interact with them. Moreover, the capacity and limits of the human being are considered and balanced with those of the system. This corresponds to a human-centred design (BERDAL et al., 2018).

According to Sivathanu and Pillai (2018), organisations will need to align their Human Resource Management (HRM) strategies and practices with Industry 4.0, including topics such as workforce employment and skills development. In this sense, certain digital skills may be needed in the future of Industry 4.0, such as problem solving, non-routine tasks and creating digital outputs (DJUMALIEVA; SLEEMAN, 2018).

5.3 UNIVERSITY-INDUSTRY

One of the main ways of transferring technology is through the interaction of universities with industries.

The role of universities in the knowledge society as producers and transmitters of knowledge is increasingly important, not only as instructors of human capital, but also in driving and creating inventions and innovations and solving social problems (DUTRÉNIT et al., 2010). This enables the use, transfer and commercialisation of knowledge as part of a new mission (SIERRA; VILLAZUL, 2018).

Based on Debackere and Veulegers (2005), university-industry collaboration refers to different types of interactions between industry and the scientific sector, which aim to exchange technology. The interaction channels involve hiring recent graduates, conferences, staff training, patents, prototypes, licensing, incubators, spin-offs, joint and cooperative ventures. However, the most common are the traditional channels, involving publications, conferences and consultancy (BARLETTA et al., 2017; LEMOS; FERRAZ, 2017; TORRES et al., 2011)

One of the main methods for generating this knowledge transfer is university *spin-offs*. In the commercial field, the term spin-off refers to the process by which a company is created from another pre-existing entity. The resulting new company is also known as a *spin-off*. The term serves to define the process itself and the outcome of that process. At the university level, one speaks of an academic or university *spin-off* when the company was established within a higher education institution, putting into practice the knowledge generated at these centres through the R&D activity of academics (MIRANDA et al., 2018).

5.4 TT FOR IMPLEMENTING TECHNOLOGIES 4.0

Technology is an essential factor in Industry 4.0. The use of new technologies requires the preparation of existing systems. This new digital transformation challenges the research and innovation landscape and requires new approaches. Platform ecosystems based on experimentation and innovation can bring stakeholders together to work on solutions and innovation opportunities for IIoT and other new technologies (VEILE et al., 2019).

According to Koch and Blind (2020), Testbeds can play a key role here in this new industrial revolution. They provide a platform for cooperation between industries, companies and academia. Thus, the implementation of new solutions and prototype development is facilitated. As bounded and spatially confined environments, testbeds enable experimentation and testing outside of real production environments (FOLSTAD, 2008; BALLON et al., 2005). They serve to think about innovations and test new applications, processes, products, services and business models to verify their usefulness and viability before taking them to market (KOCH; BLIND, 2020).

According to da Silva et al. (2019) an industry initially needs to go through internal and external TT processes to adapt to the concept of Industry 4.0. Internal TT processes are those whose industry is limited to its resources, such as knowledge transfer between experienced employees and newly hired employees. As for external processes, the industry interacts with resources from external suppliers, research institutions, government, among other organisations.

The adoption of TT approaches plays a relevant role as an enabler in the digital transformation process and is directly linked to the importance of increasing the knowledge base in I4.0 technologies. Developing specific skills for the proper use of technologies and overcoming resistance to change and lack of knowledge should be a priority objective. Internal knowledge needs to be generally enhanced, also promoting the integration between digital and traditional skills, in order to avoid that possible obstacles may arise from fragmentation (ZANGIACOMI et al., 2018).

5.5 KNOWLEDGE TRANSFER / KNOWLEDGE MANAGEMENT IN I4.0

Knowledge transfer is the process by which an individual or unit is affected by the experience of another in organizations. A production team can learn from another how to better assemble a product (ARGOTE; INGRAM, 2000). Various digital technologies have been developed to support this type of knowledge transfer in the past. The challenge of augmenting human work with digital technologies is created by effectively contributing and consuming information that gets increasingly complex, is combined from various sources and types, and constantly changes (HANNOLA et al., 2018).

In addition, new knowledge management strategies should be developed considering the dynamics resulting from the implementation of Industry 4.0. Information and communication technologies must be present at all levels of the organisation (BATZ et al., 2018).

The implementation of Industry 4.0 strategies is accelerating changes in the market and the production of material goods is acquiring several new characteristics to the service sector (BATZ et al., 2018). For this reason, qualifications and new skills are essential because companies will have to adapt to the constant technological changes. Transformation in knowledge management is key to the success of a highly innovative company (RAUCH et al., 2019).

In this new industrial revolution, the future tasks of employees require more competences. Knowledge about technologies as well as interdisciplinary knowledge must be imparted, for example through training, workshops and continuing education programmes. In addition to traditional training methods, emphasis should be placed on e-learning and scenario-based learning. Companies should also cooperate with educational institutions of all kinds in order to engage in the development and design of educational programmes tailored to the specific qualification needs of Industry 4.0 (VEILE et al., 2019).

Also, industry 4.0 relevant knowledge should be developed using research results, experiences and recommendations from branch associations and internal experiences. Sharing knowledge with, for example, research institutions, should be a reciprocal process (VEILE et al., 2019).

5.6 TT IN THE INDUSTRIAL 4.0 SCENARIO

According to Zhong et al. (2017), huge leaps have been made in technological innovation and adaptation with the movement towards the Fourth Industrial Revolution. This has attracted great attention as the rapid integration of technology into various spheres of the world has disrupted businesses, academia and government (PARK, 2018). Several studies (LANGDON et al., 2014; LASI et al.,

2014; SÁ, LEE, 2012) have observed that technology transfer plays a key role in this emerging technological paradigm. Technology transfer itself can be described as the process of transferring or disseminating technology from its creator or owner to another person (AUDRETSCH; CAIAZZA, 2016). To achieve this, several variables and actions need to occur. This usually includes a joint effort to share knowledge, skills, technologies or methods to a wide range of users who can develop and exploit the technology in new applications, materials, products, processes or services (CUNNINGHAM et al. 2019).

To adapt to the concepts of Industry 4.0, an industry initially needs to go through TT processes with suppliers from the various branches of the technology industry, mainly. Subsequently, to follow the changes already implemented in the manufacturing industry, it is necessary to use digital technologies, which are essential to facilitate the flow of information (DA SILVA et al., 2019).

According to Lee et al. (2018), various actors, including the public, private and academic sectors, will need to change the way they engage to manage the shifts in power, wealth and knowledge that can be attributed to this disruption. A key aspect of this is innovative Industry 4.0 technologies, which are being integrated across different scientific and technical disciplines, facilitating rapid advances in R&D (CARAYANNIS et al., 2017). An integral part of digital engineering, virtual and augmented reality technologies are assuming a key role in different forms of training, qualifications and knowledge transfer (SCHUMANN et al., 2015).

5.7 INNOVATION / OPEN INNOVATION IN I4.0

Industry 4.0 is highly connected to innovation. Innovation is considered an important condition for organisational building and strategic maintenance, which can increase the competitive advantage of companies, ensuring sustainability and prosperity. Collaborations and alliances have become a trend towards enabling long-term business growth. Open innovation combines internal and external resources to generate new technologies and identify new paths to market (CHESBROUGH, 2003). With open platforms, technological advancement, mobility of highly educated people and social engagement, companies can efficiently absorb knowledge resources.(YUN, LIU, 2019)

Industry 4.0 represents a shift towards an innovation-based economy with knowledge, data and IoT as central concepts. It will affect the current structure, markets and business processes of the industrial era and pave the way for a new era of digitisation, "smarter" networking of production systems and interconnected business processes (MORRAR et al., 2017).

In this new industrial revolution, traditional competitive factors, such as market share, economies of scale and access to resources, are now linked or joined to other factors such as innovation, intellectual property rights, smart technology and access to knowledge (GEIGER; SÁ, 2013). In addition,the role of the consumer in the production process has also changed: they are now co-producers. Meanwhile, radical process innovation is associated with the technological revolution; tailored production series will replace industrial or mass manufacturing facilities (BUHR, 2017).

Companies must modify business models to innovate and insert flexible value chains to increase responsiveness to changes in consumer behavior. A smart factory with intelligent production systems will meet this demand while maintaining high-quality products and services (MORRAR et al., 2017).

5.8 PREVIOUS WORK

In the systematic literature review conducted, after an intensive filtering process (Ordinate Method - see Chapter 6), 27 relevant articles were found that actually address the subjects Technology Transfer and Industry 4.0 (and its variants). These articles were thoroughly studied so as to develop this theoretical referential and ground the research in question.

Table 10 summarises what each of these papers studied and their main objectives.

Table. 10 - Objectives of each article

Article	Aim of the study
Karre et al. (2017)	The authors reviewed the classification of the *TU Graz LeanLab* according to the morphology of teaching factories, summarised the upcoming challenges in manufacturing in the context of Industry 4.0 and created an implementation roadmap for the specific case.
da Silva et al. (2019)	This study contextualizes TT in the Industrial 4.0 scenario of the supply chain, focusing on the stages of supply, manufacturing industry and final consumer. The results infer that, in the Industrial 4.0 scenario, the supply chain will undergo changes, such as real-time visibility throughout the supply chain, continuous collaboration between the stages of the chain, among other significant changes.
Al-Saeed et	This paper presents a BIM Digital Objects (BDO) framework to automate building

al. (2020)	product manufacturers' processes and increase lean manufacturing, through a case study of a knowledge transfer project.
Ansari et al. (2020)	The main objective of this paper is to establish a methodology for the optimal selection of an employee competent in I4.0 requirements, corresponding to the characteristics of the task and learning requirements, including knowledge, skills and competencies.
Kruger, Steyn (2020)	Thus, the aim of this study is to demonstrate the role of innovation spaces in knowledge-rich research environments, such as universities, and how they enhance entrepreneurial skills to create high-value ventures. This can be attributed to the development of new generation technologies arising from the Fourth Industrial Revolution.
Mahmood, Mubarik (2020)	This study examines the role of intellectual capital (IC) - human capital, structural capital and relational capital - in balancing innovation and exploitation activities in Industry 4.0.
Michna, Kmieciak (2020)	The aim of this paper is to investigate whether open-minded culture, knowledge sharing and financial performance have a significant impact on the willingness to implement Industry 4.0 in SMEs.
Koch, Blind (2020)	In this paper, the authors conduct an exploratory multiple case study on testbeds in the context of Industry 4.0 to find out how they impact standardisation processes, principles and outcomes.
Barbosa et al. (2020)	This paper emphasizes the proposal of a customized roadmap that manages the knowledge related to automation skills to be applied during the concurrent engineering phase. The approach of this R&D roadmap is a particular guideline in a structured way that uses the PDCA method concepts integrated with other quality tools and know-how in robotics, for managing this specific knowledge.
Tirto et al. (2020)	The aim of this article is to consider the technology transfer and education issues of ICT support for Industry 4.0 innovation networks. The study is based on a broad understanding of technology transfer as an exchange of technology, technology knowledge between individuals, companies, universities, research centres and government structures at all levels. The proposed research idea is also based on the concept of integrated technology and the concept of promising needs (innovation).
Gjeldum et al. (2016)	The aim of the Innovative *Smart Enterprise* project is to establish a special learning environment in a laboratory like *Lean Learning Factory*, i.e. the simulation of a real factory by means of specialised equipment. Adapted and implemented methods and tools in the design and analysis process for optimising this hybrid assembly line would be scaled and adjusted for use in industry as part of the knowledge transfer from university to companies.
Fu et al. (2019)	In this paper, the authors presented a new multi-scale knowledge transfer method called MSKT. This method transfers multi-scale knowledge from the teacher network to the student network, making the student suitable for IoT applications in Industry 4.0.
Ituarte et al. (2018)	Additive manufacturing requires a systemic approach to assist industry in researching technology applicability. To this end, the aim of this research is to help manufacturing business leaders decide whether digitised manufacturing based on additive manufacturing is suitable for engineering applications and help them plan technology transfer decisions.
Cianfanelli et al. (2018)	This paper presents the results of a new concept of reverse engineering of the product applied to the practical craft-based manufacturing process. It is a partnership between University and Company, which the authors call *Joint Labs,* focusing on 3D scanning, re-modelling and 3D design.
Liboni et al. (2019)	The aim of this article is to address the potential impacts of Industry 4.0 on human resource management - with a particular focus on employment, job profile and qualification and skills requirements in the workforce - that may have implications for supply chain management. Consequently, exploratory relationships between Industry 4.0, HRM and SCM are presented based on a systematic review.
Rauch et al. (2020)	This article presents a systematic literature review of the anthropocentric perspective of production before and after Industry 4.0. The authors identified core research clusters in relation to traditional anthropocentric production systems (APS) and anthropocentric cybernetic physical production systems. By comparing the two perspectives, they were able to analyse new emerging paradigms in anthropocentric production caused by Industry 4.0. Furthermore, they made a prediction of the future

	role of the human operator, their required knowledge and capabilities and how assistance systems support the Operator 4.0.
Berdal et al. (2018)	The paper focuses on the design of self-organised production systems that cooperate with humans. A literature review is provided based on two views dealing with such design: a technical and a human-machine system. Limits and advantages of both views are presented.
Posselt et al. (2016)	This study focuses on the approach of teaching factories for Industry 4.0. A central approach is to pursue the new interactive knowledge transfer through the multisensory approach combined with innovative feedback processes, enabling a learning process with all human senses.
Feng et al. (2017)	This paper prescribes knowledge management to address the lack of mechanisms to integrate, share and update domain-specific knowledge in smart manufacturing. Aspects of knowledge constructs include conceptual design, detailed design, process planning, material ownership, production and inspection.
Yun, Liu (2019)	This article explores how sustainability can be achieved through open innovation in the 4th industrial revolution. The authors have developed a framework to understand the micro and macro dynamics of open innovation with a quadruple helix model for social, environmental, economic, cultural, political and knowledge sustainability.
Yun et al. (2018)	With the advent of the Fourth Industrial Revolution, the Korean industrial environment has changed rapidly. As a result, various industries are facing different innovation conditions. This study used parameters related to patents originating from the automotive, robotics and aerospace fields. The patents subject to analysis were those registered in Korea from 2001 to 2014 and included those applied for, jointly applied for or registered by Korean companies. This study analysed the differences caused by the effect of open innovation in the three industries.
Hannola et al. (2018)	This article proposes a *framework* for empowering workers in industrial production environments with digitally facilitated knowledge management processes. The framework explores four concrete facets of digital advances that apply to a wide range of knowledge processes and production strategies in manufacturing firms. Each of these advances is capable of supporting a specific facet of individual workers' knowledge management processes: knowledge transfer, discovery, acquisition and sharing.
Schumann et al. (2015)	This paper examines a means by which digital engineering, virtual reality and augmented reality technologies can support the creation of sustainable smart manufacturing and logistics processes, as well as on-the-job training and skills and knowledge transfer.
Sastoque et al. (2019)	In this article a study for the creation of new Advanced Manufacturing centres is presented, centres whose activities are focused on Technology Readiness Levels (TRL) 5-7. The foundations of the sector knowledge and the difficulties pointed out are established so that it can guide similar initiatives.
Garbellano, da Veiga (2019)	This article aims to focus on how Industry 4.0 technology transfer has been implemented by innovative Italian small and medium-sized enterprises (SMEs).
Ilie and Gheorghe (2014)	Nanotechnologies can constantly underpin innovation and technology transfer, smart products and technologies, creating jobs and helping economic growth and a sustainable economy. The aim of this article is to highlight elements of technology transfer to enhance the smart manufacturing process.
Tourk, Marsh (2016)	This article examines the contours shaping the New Industrial Revolution and looks at the challenges China faces in its quest to achieve technological parity with advanced nations, with the aid of R&D, published papers, patents and innovation centres.

Source: Own Authors (2021)

All these papers have studied TT and I4.0, but in different ways. Some of them focused on knowledge transfer, others on TT for implementing 4.0 technologies, others still on teaching technologies relevant to I4.0. Still, it is noticeable that some papers present I4.0 as advanced manufacturing, factories of the future, smart factories, and so on.

In this sense, a chart was conducted to demonstrate in a simplified manner, which relationship authors present between Industry 4.0 and Technology Transfer. This classification is presented in Chart 11. An article may be inserted in more than one category.

Box. 11 - Relationship between I4.0 and TT presented in each article

Author (year)	Learning Factories	Human-Machine Interaction	University - Industry	TT for 4.0 technologies implementation	Knowledge Transfer / Knowledge Management in I4.0	TT in the industrial 4.0 scenario	Innovation / Open Innovation in I4.0
Karre et al. (2017)	✓		✓		✓		
da Silva et al. (2019)						✓	
Al-Saeed et al. (2020)				✓			
Ansari et al. (2020)		✓			✓		
Kruger, Steyn (2019)			✓			✓	✓
Mahmood, Mubarik (2020)					✓		✓
Michna, Kmieciak (2020)				✓	✓		
Koch, Blind (2020)				✓			✓
Barbosa et al. (2020)			✓		✓		
Tirto et al. (2020)		✓					✓
Gjeldum et al. (2016)	✓				✓		
Fu et al. (2020)				✓	✓	✓	
Ituarte et al. (2018)						✓	
Cianfanelli et al. (2019)			✓		✓		
Liboni et al. (2019)		✓			✓		
Rauch et al. (2019)		✓			✓		
Berdal et al. (2019)		✓					
Posselt et al. (2016)	✓						
Feng et al. (2017)					✓		
Yun, Liu (2019)							✓

Yun et al. (2018)							✓
Hannola et al. (2018)		✓			✓		
Schumann et al. (2015)						✓	
Sastoque et al. (2019)			✓	✓			
Garbellano, Da Veiga (2019)				✓			
Ilie, Gheorghe (2014)				✓		✓	
Tourk, Marsh (2016)						✓	

Source: Own Authors (2021)

Knowing how this relationship occurs is very important for the understanding of the presence of Technology Transfer in Industry 4.0. Researchers and organisations can benefit from this categorisation, as it shows different ways in which these two themes are influenced by each other.

Knowledge Transfer / Knowledge Management in I4.0 is the category that presents a higher concentration of articles, with 12 in total. Then, the categories *TT for implementation of technologies 4.0* and *TT in the industrial scenario 4.0* are the ones that have more studies, each with seven articles.

There are several gaps and suggestions for future work in these articles studied. Some pertinent suggestions proposed by these authors are presented below.

In the work of Liboni et al. (2019), the authors realized that there are also opportunities for studies that analyse other HRM dimensions, as the focus has been on training and development, leadership and teamwork, and communication. According to them, it is necessary to study new models of selection and recruitment, as well as performance and new competencies.

The limitation of the study of da Silva et al. (2019), according to the authors, is that it is restricted to the technological aspects related to Industry 4.0, not addressing other aspects that should be discussed for the industries maturation. In this sense, for further studies, the authors suggest that practical contributions of supply chain management in Industry 4.0, as well as applied approaches related to TT in the supply chain context in the Industrial 4.0 scenario, can be developed.

For Liboni et al. (2019), there is a great opportunity to discuss new demands on skills and education, to create a broader and better structured knowledge of the basic concepts related to this new industrial revolution. Research on new jobs and new skill requirements will be needed, and the role of universities and education from this perspective should be discussed. Research focusing on how the labour market will be affected by the implementation of collaborative robot partnerships is also on the agenda for future studies, as well as discussions about technological changes resulting in inequality and a possible expansion of precarious jobs.

Also according to the authors, there is also an opportunity to study the differences between countries, as the level of automation of each nation naturally adapts a faster performance in organizational transformations towards Industry 4.0. Moreover, the social and technological impacts of this industrial revolution are felt differently in each nation, and this should present a rich research topic (LIBONI et al., 2019).

Finally, Feng et al. (2017) point out that future work needs to develop more case studies to demonstrate the reuse of such knowledge constructs and the usefulness of a smart manufacturing knowledge base. To realize a wide dissemination of knowledge management methodology, it is crucial to store a wide variety and large number of knowledge artifacts.

5.9 THE INFLUENCE OF TECHNOLOGY TRANSFER IN INDUSTRY 4.0 : CONCEPTUAL FRAMEWORK

This chapter's intent was to describe the existing relation between Industry 4.0 and Technology Transfer. All articles found in the Systematic Literature Review that presented these two themes as the prime focus were studied.

Based on all that was studied and analysed in chapters 2, 3, 4 and in this chapter, it is possible to reach the conclusion that Technology Transfer influences Industry 4.0. And also, vice-versa.

Da Silva et al. (2019) states that for the implementation of an Industry 4.0 an evolutionary process is required, and for this process to occur, the presence of Technology Transfer is essential. Only the 4.0 concepts do not guarantee the implementation of this new industrial revolution.

Given the above in this review, the following theory was raised:

The set of Technology Transfer with the Concepts and Technologies 4.0, applied in five dimensions (People, Process, Technology, Strategy and Organisation, and Integration), facilitate the implementation process of Industry 4.0. In this sense, Figure 20 presents the TT-I4.0 Conceptual Framework, which illustrates this new concept.

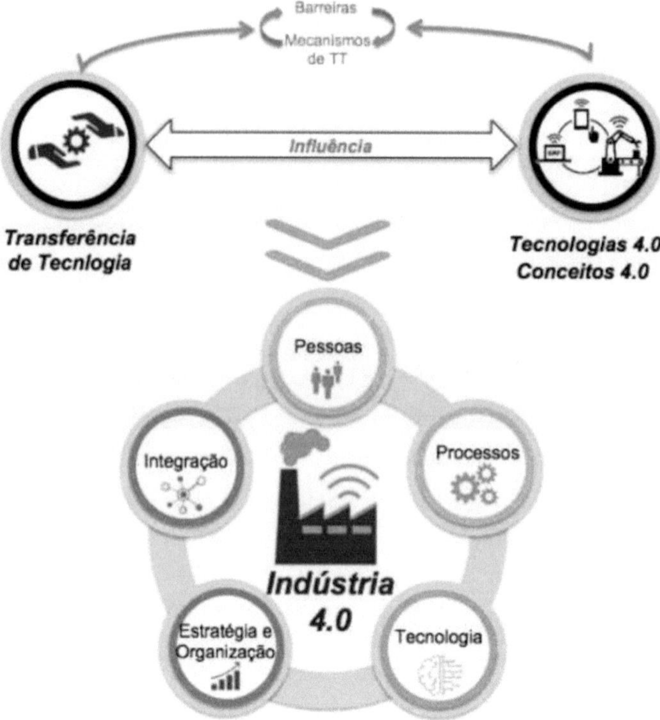

Figure 20 - TT-I4.0 Conceptual Framework
Source: Own authorship (2021)

TT mechanisms can minimise the barriers encountered in its implementation, facilitating the process. Likewise, existing I4.0 technologies and principles mean that new changes need to be adopted for an effective and current Technology Transfer.

Technology Transfer with 4.0 concepts and technologies act directly on the five dimensions:

(1) *People:* This refers to Human-Machine interaction, and all the training, knowledge and skills development required for I4.0 (SCHUMACHER et al.,2016; SJÖDIN et al., 2018; BEER et al., 2017; SECUNDO et al., 2016; MENDOZA AND SANCHEZ, 2018);

(2) *Processes:* Refers to the processes required in an I4.0 for product and/or service development (AKDIL et al., 2018; SJÖDIN et al., 2018; JOCHEM et al., 2011);

(3) *Technology:* Refers to the technologies (materials and knowledge) that should compose the transformation process of the product and/or service (SCHUMACHER et al., 2016; SJÖDIN et al., 2018; BEER et al., 2017; SECUNDO et al., 2016; MENDOZA AND SANCHEZ, 2018);
(4) *Strategy and organisation: This* refers to the fit with new Business Models and the organisational culture in the company (SCHUMACHER et al., 2016; SCHUH et al., 2017; AKDIL et al., 2018; MENDOZA AND SANCHEZ, 2018);
(5) *Integration: This* refers to the way systems are interconnected in Industry and Supply Chain 4.0 (SCHUH et al., 2017; BEER et al., 2017; JOCHEM et al., 2011).

In this way, Technology Transfer becomes an ally to 4.0 Concepts and Technologies for the evolutionary implementation of Industry 4.0 in organisations.

The works reported in this chapter presented the importance of TT in I.40. However, the gap found was that none of these works presented what really is this existing influence and how it can be analysed. Thus, it was performed this theoretical analysis that arrived at the approach presented in Figure 20.

To prove and apply this raised theory, the next chapter discusses the methodological procedures used to develop this research.

6 METHODOLOGY

To achieve the objectives proposed in this work, this chapter presents the methodological procedures that were adopted. The chapter is subdivided into the following sections: Systematized Bibliographical Review and Model Elaboration. The first section describes how the bibliographical portfolio was obtained. In the second section, it is described how the Maturity Model was built in stages.

6.1 SYSTEMATISED LITERATURE REVIEW (SBR)

Obtaining the bibliographic portfolio was based on the search through a systematized literature review (RBS), using the *Methodi Ordinatio* methodology of Pagani, Kovaleski and Resende (2015). This methodology orders articles based on criteria such as the impact factor of the journals to which the articles were published, the year of publication and number of citations. For the application of this methodology, the steps that must be followed are represented in Figure 21.

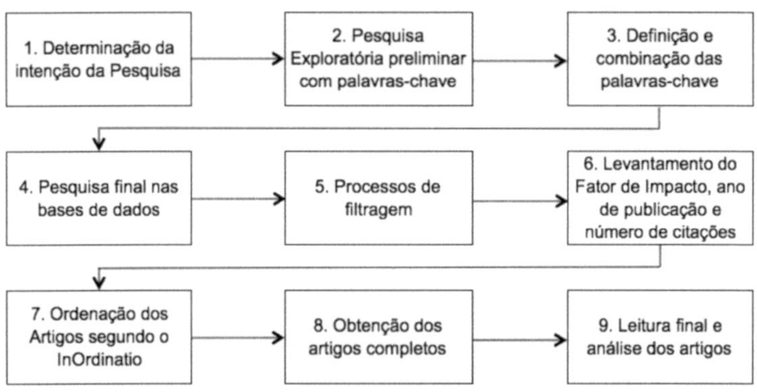

Figure 21 - Stages of Mehodi Ordinatio
Source: Adapted from Pagani, Kovaleski and Resende (2015)

After fulfilling the 9 steps of this method, 314 articles were studied to extract the necessary information for the composition of this work. Thus, it was performed a content analysis to achieve the research objectives. All these steps are detailed in Appendix A.

6.2 ELABORATION OF THE I4.0-TT MATURITY MODEL

To achieve the objectives proposed in this work, a model was developed where it is possible to analyse the I4.0 maturity level of an organisation, based on TT and 4.0 concepts and technologies.

This model will prove and apply the conceptual *framework* proposed in Section 5.4. It states that the influence of Technology Transfer applied to 4.0 concepts and technologies facilitates the implementation of Industry 4.0. For simplicity, these 4.0 concepts and technologies will be referred to as *Building Blocks* 4.0 (BB4.0) in this model as they are the necessary components for Industry 4.0.

This maturity model is based on the influence between the factors (TT and BB4.0). Thus, it will be possible to verify whether these factors infer each other or not.

To build this model, 7 steps are required, presented in Figure 22 and described below.

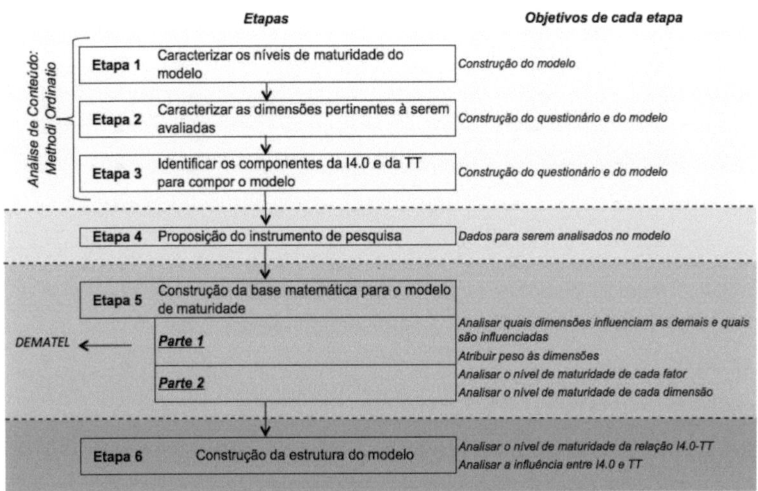

Figure 22 - Stages for the development of the I4.0-TT Maturity Model

Source: Author (2021)

6.2.1 Step 1: Characterise the model's maturity levels

The analysis of the models found in the literature allowed the maturity levels determined by each researcher to be visualized. These models present an average of five levels, and when each maturity level is reached by the industry, it means that it will be evolving in the maturity of its processes.

In this work it was determined that the model will present four levels of maturity for *Building Blocks* 4.0 and four levels of maturity for Technology Transfer.

The levels of maturity in the model are described in Table 12:

Box. 12 - Maturity model levels

Maturity Level	Description	Authors
Level 1 - *Initial*	The factor analysed is at a critical level, it needs to advance a lot.	Ganzarain and Errasti (2016); Kreiling and Bounfour (2020); Wagire *et al.* (2020).
Level 2 - *Defined*	The existence of some components of the factor analysed varies from an alert to an acceptable state.	Jochem *et al.* (2011); Ganzarain and Errasti (2016); Secundo *et al.* (2016); Facchini *et al.* (2020); Kreiling and Bounfour (2020).
Level 3 - *Managed*	The components of the factors analysed are measured and controlled, is at an optimal level.	Jochem *et al.* (2011); Secundo et al. (2016); Ganzarain and Errasti (2016); Gokalp et *al.* (2017); Beer *et al.* (2017); Facchini *et al.* (2020); Kreiling and Bounfour (2020).
Level 4 - *Sustained Maturity*	It is the ideal level, where all components are integrated, measured and controlled. At this stage, the focus is on Continuous Improvement.	Secundo *et al.* (2016); Beer et al. (2017); Akdil *et al.* (2018); Facchini *et al.* (2020).

Source: Own Authors (2021)

6.2.2 Step 2: Characterise the relevant dimensions to be assessed

At this stage it is necessary to determine the dimensions that will be analysed in the model. These dimensions were determined based on the existing I4.0 and TT models and on the literature review, as presented in Section 5.4.

As there are two factors to be measured, the dimensions chosen need to fit and be related to both I4.0 and TT. As presented in Section 5.4, five dimensions were determined to be analysed in this model. These five dimensions were sub-divided into Technology Transfer Dimensions and *Building Blocks* 4.0 Dimensions, totalling 10 dimensions. Table 13 presents these dimensions, sub-dimensions and their main characteristics.

Box. 13 - Dimensions of the maturity model

Dimensions	Dimensions TT	Dimensions BB4.0	Authors
People	*Human Resources* It refers to all factors related to human resources, such as training, skills development, among others.	*Human-Machine Interface* It refers to man's existing relationship with technology.	Schumacher et al. (2016); Sjödin et al. (2018); Beer et al. (2017); Secundo et al. (2016); Mendoza and Sanchez (2018)

Process	*TT Process* Refers to the resources required to achieve a TT.	*I4.0 Process* Refers to the resources needed in an I4.0.	Akdil et al. (2018); Sjödin et al. (2018); Jochem et al. (2011)
Technology	*Dissemination of Knowledge* It refers to the ways in which technology and knowledge transfer is disseminated in the organisation.	*Technologies 4.0* Refers to the technologies that should be part of the transformation process of the product / service.	Schumacher et al. (2016); Sjödin et al. (2018); Beer et al. (2017); Secundo et al. (2016); Mendoza and Sanchez (2018)
Strategy and Organisation	*Organisational Culture* It refers to the existing culture in the organisation in relation to TT.	*Business Models* It refers to the adaptation to new business models.	Schumacher et al. (2016) ; Schuh et al. (2017); Akdil et al. (2018); Mendoza and Sanchez (2018)
Integration	*TT Modes* It refers to the ways in which TT exists in the organisation, how technology is transferred from one place to another.	*Systems Integration* It refers to the way systems are interconnected in Industry and Supply Chain 4.0.	Schuh et al. (2017); Beer et al. (2017); Jochem et al. (2011)

Source: Own Authors (2021)

Based on these 10 dimensions, the research instrument will be built.

6.2.3 Step 3: Identify the components of I4.0 and TT to compose the model

The next step is to identify all the Industry 4.0 (BB4.0) and Technology Transfer components required to make up the proposed model.

All these components were selected via the Systematic Literature Review conducted. The factors that authors point out as being essential to the functioning of an Industry 4.0 and/or Technology Transfer were tabulated. This means that with the analysis of the presence or absence of these components, it will be possible to measure the maturity level of TT and BB4.0. In this way, these components will be analyzed through the questionnaire proposed in Step 4. Table 14 presents all these components of BB4.0 and TT.

Box. 14 - Components of the maturity model

Components			Authors
Industry 4.0	1	Smart Factory	Qin et al. (2016); Pereira and Romero (2017); Radziwon et al. (2014); Kagermann et al. (2013).
	2	Smart Products	Qin et al. (2016); Pereira and Romero (2017); Kagermann et al. (2013).
	3	Business Model	Qin et al. (2016); Pereira and Romero (2017)
	4	Staff Competences	Qin et al. (2016); Pereira and Romero (2017); Stock et al. (2018).
	5	Employee autonomy	Qin et al. (2016); Pereira and Romero (2017)
	6	Horizontal Integration	Wang et al. (2016); Stock et al. (2018)
	7	Vertical Integration	Wang et al. (2016); Pereira and Romero (2017)
	8	End-to-end engineering integration	Wang et al. (2016); Stock et al. (2018)
	9	Human-Machine Interface	Bahrin et al. (2016); Liboni et al. (2019); Thramboulidis, Christoulakis (2016); Pfeiffer (2016); Sivathanu and

			Pillai (2018)
	10	Interoperability	Carvalho et al. (2018); Qin et al. (2016); Persson (2016).
	11	Decentralisation	Carvalho et al. (2018); Kamble et al. (2018); Persson (2016).
	12	Virtualisation	Carvalho et al. (2018); Kamble et al. (2018); Persson (2016).
	13	Real-time work	Carvalho et al. (2018); Kamble et al. (2018); P Persson (2016).
	14	Modularity	Carvalho et al. (2018); Qin et al. (2016); Persson (2016).
	15	Service orientation	Carvalho et al. (2018); Kamble et al. (2018); Persson (2016).
	16	Connectivity	Kumar and Nayyar (2020)
	17	Control systems	Kumar and Nayyar (2020)
	18	Smart Sensors	Kumar and Nayyar (2020)
	19	Cyber and Physical System	Varghese, Tandur (2014); Lee et al. (2015); de Silva, de Silva (2016); Zhou et al. (2016); Mladineo et al. (2017)
	20	*Cloud* computing	Wang et al. (2014); Branger, Pang (2015); Thames, Schaefer (2016); Moghaddam, Nof (2017).
	21	Industrial Internet of Things (IIOT)	Sisinni et al. (2018); Wong, Kim (2017); Thramboulidis, Christoulakis (2016)
	22	Augmented Reality (AR)	Lee et al. (2011)
	23	*Big Data Analytics*	Babiceanu, Seker (2016); Denkena et al. (2014); Kushiro et al. (2014); Rago (2015)
	24	*Cyber* Security	Thames, Schaefer (2017)
	25	Autonomous Robots	Moeuf et al. (2018); Pei et al. (2017)
	26	*Additive manufacturing / 3D* Printing	Chen, Lin (2017); Jian et al. (2017)
	27	Simulations	Rüßmann et al. (2015); Bahrin et al. (2016).
	28	Existence of central coordination for I4.0	Schumacher et al. (2016)
	29	Artificial Intelligence	Esmaeilian et al. (2016)
	30	Performance indicators	Pereira and Romero (2017); Wang et al. (2016); Esmaeilian et al. (2016).
	31	Digitised communication	Esmaeilian et al. (2016)
	32	Marketing 4.0	Jara et al. (2012)
Technology Transfer	33	University-Company Interaction	Stephan (2001); Debackere, Veugelers (2005); Cormican and O'Connor (2009); Dutrénit et al. (2010); Sierra, Villazul (2018).
	34	Knowledge sharing	Argote and Ingram (2000)
	35	Vertical technology transfer	Ockwell et al. (2008); Park, Lee (2011)
	36	Horizontal technology transfer	Ockwell et al. (2008); Park, Lee (2011)
	37	Technology provider	Mendoza and Sanchez (2018)
	38	Receiver of the technology	Mendoza and Sanchez (2018)
	39	Intermediary agent	Mendoza and Sanchez (2018)
	40	Buyer-Supplier Interaction	Ockwell et al. (2008); Shujing (2012); Pueyo et al. (2011)
	41	Training	Chege et al. (2019); Morgera and Ntona (2018); Lema et al. (2018); Eitzel et al. (2018); Bruckman et al. (2018).
	42	Consultancies	Debackere, Veugelers (2005); Khalozadeh et al., (2011); Trencher et al. (2013); Vac and Fitiu (2017); Bliznets et al. (2018)
	43	Networking and resource sharing	Bozeman (2000); Canestrino (2009).
	44	Exchange of staff, researchers or professionals	Manyuchi (2017); Basu (2018); Bliznets et al. (2018); Lema et al. (2018); Sarkodie and Strezov (2019).
	45	Internship	Cohen et al. (2002).
	46	Inter-institutional networks	Mehta et al. (2016); Morgera, Ntona (2018); Bruckman

		et al. (2018); Li et al. (2018); Sarkodie, Strezov (2019)
47	Research and Development	Urban et al. (2015); Mehta et al. (2016); Aggarwal and Aggarwal (2017); Li et al. (2018); Bliznets et al. (2018)
48	*Spin-offs*	Gilsing et al. (2011); Looy et al. (2011); Sánchez et al. (2012); Trencher et al. (2013); Vac, Fitiu (2017)
49	*Joint Venture*	Khalozadeh et al. (2011); Mohamed et al. (2012); Urban et al. (2015); Mehta et al. (2016); Bliznets et al. (2018)
50	Publications	Ferdows (2006); Gilsing et al. (2011); Trencher et al. (2013); Pinard (2013); Escalante et al. (2013).
51	Seminars / Workshops	Doukas et al. (2012); Trencher et al. (2013); Pinard (2013); Belmonte et al. (2015); Eitzel et al. (2018).
52	Patents / Intellectual Property	Park et al. (2013); Trencher et al. (2013); Shi and Lai (2013); Bliznets et al. (2018); Ferreira et al. (2018).
53	Skills development	Lanza et al. (2015); Wang et al. (2015); Morell, Trucco (2012)
54	Man-Machine Relationship	Bahrin et al. (2016); Liboni et al. (2019); Thramboulidis, Christoulakis (2016); Pfeiffer (2016); Sivathanu, Pillai (2018)
55	Skills and methods management	Schumacher et al. (2016)
56	Innovation-open	Schumacher et al. (2016)
57	Interdisciplinary teams	Palacios-Marqués et al. (2013)

Source: Own Authors (2021)

6.2.4 Stage 4: Proposing the research instrument

At this stage, the research instrument is proposed. For the model composition it is necessary to collect information from industries to analyze their maturity levels in the context of I4.0 and TT. The questionnaire was chosen as instrument for data collection because it has the following advantages: freedom of time for respondents, replicability, accuracy in the answers, practicality and the anonymity of the participating industry.

The questionnaire was developed from the selected components of I4.0 and TT (Step 3), where they were framed in the defined dimensions (Step 2). With this it is possible to analyze the maturity of each dimension and each factor (BB4.0 and TT).

The theoretical constructs used in this questionnaire were based on previous research. The survey instrument was pre-tested with 3 subject matter experts. Their *feedbacks* were used to improve the format and clarity of the survey.

There were two questionnaires that the respondents answered, being:

(1) Questionnaire 1 - The respondent should answer a questionnaire concerning the influence that each dimension has on each other. The question is the following: What influence does dimension y have on dimension x? Thus, the company representative must indicate the level of influence between each dimension on a scale of 0 to 4: 0

(no influence); 1 (low influence); 2 (medium influence); 3 (high influence); 4 (very high influence). This questionnaire is presented in Appendix B.

(2) Questionnaire 2 - In this questionnaire are the questions referring to all the components selected in Step 3. There is a total of 60 questions in the questionnaire, being 30 for each factor (I4.0 and TT). These questions were categorized in ten dimensions: Human Resources, TT Process, Knowledge Dissemination, Organizational Culture, TT Modes, Human-Machine Interface, I4.0 Process, Technologies, Business Models and System Integration. All items are perceptual measurement using a 5-point *Likert* scale: 0 (does not occur); 1 (very rarely); 2 (occurs to a small extent); 3 (occurs somewhat frequently); 4 (occurs very frequently / always). This questionnaire is presented in Appendix C.

Next, Stage 5 of the work will be presented, that is, the mathematical basis of the I4.0-TT Maturity Model.

6.2.5 Stage 5: Construction of the mathematical basis for the maturity model

The mathematical basis of the I4.0-TT Maturity Model is divided into two parts. The first part aims to analyse which TT and BB4.0 dimensions influence the others, and which dimensions are influenced. In addition, weights will also be assigned to these dimensions at this stage, according to the overall impact given in relation to the others. The second part aims to analyze the maturity level of each dimension and factor.

6.2.5.1 First part

The influence between TT and BB4.0 will be analysed. Thus, the model is a function of these two factors. Since each of these factors has five dimensions that define them, these also have influences on each other. Therefore, the following mathematical equation can be written for the model:

MODEL = f (TT, BB4.0);
MODEL = f (RH, PTT, DC, CO, MTT, IHM, PI4.0, T4.0, MN, IS);
RH= f (PTT, DC, CO, MTT, IHM, PI4.0, T4.0, MN, IS);
PTT= f (RH, DC, CO, MTT, IHM, PI4.0, T4.0, MN, IS);
DC= f (RH, PTT, CO, MTT, IHM, PI4.0, T4.0, MN, IS);

CO= f (RH, PTT, DC, MTT, IHM, PI4.0, T4.0, MN, IS);
MTT= f (RH, PTT, DC, MTT, IHM, PI4.0, T4.0, MN, IS);
HMI= f (RH, PTT, DC, CO, MTT, PI4.0, T4.0, MN, IS);
PI4.0= f (RH, PTT, DC, CO, MTT, HMI, T4.0, MN, IS);
T4.0= f (RH, PTT, DC, CO, MTT, HMI, PI4.0, MN, IS);
MN= f (RH, PTT, DC, CO, MTT, IHM, PI4.0, T4.0, IS);
IS= f (RH, PTT, DC, CO, MTT, IHM, PI4.0, T4.0, MN).

where,

HR= Human Resources; PTT= TT Process; DC= Knowledge Dissemination; CO= Organisational Culture; MTT= Modes of TT; HMI= Human Machine Interface; PI4.0=Process of I4.0; T4.0= Technologies 4.0; MN= Business Models; IS= Systems Integration.

This means that the proposed model is a function of all dimensions and that each dimension is a function of the others: they are all interconnected, and depend on or have influence over the others. To better visualise it, the following scheme was developed (Figure 23).

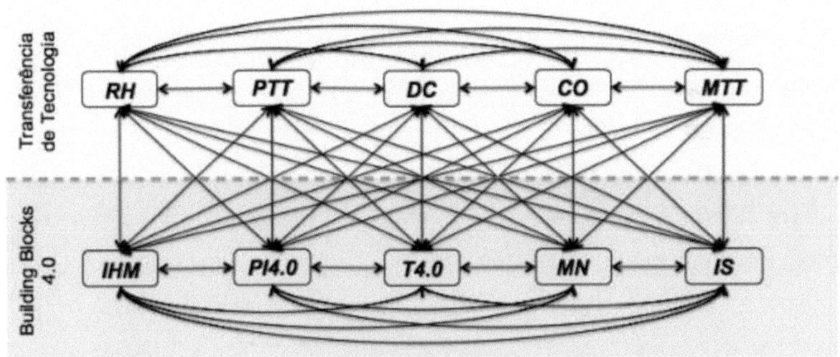

Figure 23 - Influence diagram between TT and BB4.0 dimensions
Source: Author (2021)

Its application consists of experts indicating the level of influence between each dimension (Questionnaire 1), on a scale of 0 to 4: 0 (no influence); 1 (low influence); 2 (medium influence); 3 (high influence); 4 (very high influence). Next, the DEMATEL method is applied.

The DEMATEL method makes it possible to analyse which of these dimensions are responsible for influencing the others and which are the influenced dimensions. By applying this method, it is possible to obtain the weight for each of the dimensions.

The multicriteria method DEMATEL was used in this model since it is the main MCDM (*Multiple-Criteria Decision Method*) model capable of testing the influence between factors. DEMATEL is an extended method for building a structural model to analyse the influence relationship between complex criteria (CHANG et al., 2011).

The DEMATEL method

The DEMATEL - Decision Making Trial and Evaluation Laboratory approach was developed by Gabus and Fontela (1972) and aims at analysing a set of components, factors, alternatives or criteria, capable of exerting influence on each other not always in a reciprocal manner.

Based on Wu and Lee (2007), this method by Gabus and Fontela (1972) can be summarised as follows:

Step 1. Construct the cross-relationship matrix A expressing the level of influence that element *i* in the row of the matrix exerts over element *j* in the column of the matrix, where □□□ is the influence that element i exerts over element j. For this, a scale of comparison of influence levels must be established. On a scale of 0 to 4, 0 means no influence, 1 little influence, 2 medium influence, 3 high influence and 4 very high influence.

Step 2. construct the direct relationship matrix D by calibrating the previous matrix according to Equation 1 and Equation 2.

$$D = A \times S$$

Equation (1)

where:

$$s = \max\left(\max_{1 \le i \le n} \sum_{j=1}^{n} a_{ij} , \max_{i \le j \le n} \sum_{i=1}^{n} a_{ij} \right)^{-1}$$

Equation (2)

Step 3. Construct the total relationship matrix *T* according to Equation 3, where, *I* is the identity matrix and (□ - □)-1 an inverse matrix.

$$T = D(I - D)^{-1}$$

Equation (3)

Step 4. Calculate the sum of each row □□ of matrix T (Equation 4) and the sum of each column □□ of matrix T (Equation 5).

$$r_i = \sum_{j=1}^{n} t_{ij}$$

Equation (4)

$$c_i = \sum_{i=1}^{n} t_{ij}$$

Equation (5)

The sum of the row for each element □□ represents the total impact each element has on the set of elements studied. The sum of the column of each element □□ represents the total impact received by each element in the set of elements studied.

One should also calculate for each element the sum □□ + □□, which represents the total involvement that each element has in the set of elements studied and the difference □□ - □□, which represents the net effect that each element has in the set of elements studied.

The DEMATEL approach classifies the components studied into 2 groups: the influenced, which have negative r-c and the influencers, which have positive r-c. This is because the influencers exert a degree of impact greater than the degree of impact they receive within the studied set. And those influenced receive a degree of impact greater than the degree of impact they exert on the other components within the studied set.

6.2.5.2 Part Two

In the second part of the model, a manager responsible for each company must answer Questionnaire 2 (Appendix C) prepared in Step 4. After its application and the collection of replies, it is possible to conclude the mathematical development of the model proposed in this research.

For each dimension a weight (*w*) is attributed through the DEMATEL method presented, in relation to its total impact on the set of dimensions (*ri*). Initially, the maturity of each dimension, for each of the factors (BB4.0 and TT), is extracted via Equation 6:

$$M_d = \frac{V_d}{Ni}$$

Equation (6)

whereby:

$$V_d = \sum_{i=1}^{n} Q * w$$

Equation (7)

where:

V_d= Value of the dimension;

Ni= Number of items measured in the dimension;

Q = the answer to the question;

w = dimension weight (normalised *ri*);

M_d= Maturity of the dimension.

It is worth noting that the application of Equation 6 only results in the value of each dimension in a single analyzed company. Therefore, the procedure must be repeated for all the companies studied, for all the dimensions and for the two factors. Finally, the sum of all Values of the dimensions (V_d) results in the Maturity of each Factor (TT and I4.0) in each company, using Equation 8.

$$M_f = \sum_{i=1}^{n} V_d$$

Equation (8)

where:

M_f= Factor Maturity;

V_d= Value of the Dimension;

With the results of Equation 8, it is possible to analyze what is the Maturity Level of each factor (TT and BB4.0) in each company studied. And thus, obtain the Industry 4.0 Maturity level, which will be presented in Step 6.

6.2.6 Stage 6: Building the I4.0-TT Maturity Model structure

Finally, the structural construction of the I4.0-TT Maturity Model was carried out with the objective of facilitating the analysis of the positioning of companies in relation to the maturity of Industry 4.0, *Building Blocks* 4.0 and Technology Transfer.

In this model, the horizontal axis x is assumed as *Building Blocks* 4.0 indicator and the vertical axis y as Technology Transfer indicator. It is assumed that these two factors are dependent on each other.

To determine the maximum value of the x and y axes, it is adopted that the influence between the dimensions have the maximum value. Thus, the weight (w) of each dimension is 0.1. Subsequently, it is adopted that all questions of the dimensions obtained the maximum score (score 4 in all questions of Questionnaire 2). Performing the whole procedure described in the Second Part of Step 5, the Maturity of each Factor (M_f) is equal to 12. Thus, it is assumed that the scale of the axes is from 0 to 12.

In each of the axes, the Maturity Levels determined in Step 1 were placed. For each level, the following value was assumed:

- Level 1: $0.0 \leq M_f < 4{,}0$;
- Level 2: $4.0 \leq M_f < 8{,}0$;
- Level 3: $8.0 \leq M_f < 12{,}0$;
- Level 4: $M_f = 12{,}0$.

Thus, with the intersection of the values of the Maturity Levels of each company, the x-axis (BB4.0) and the y-axis (TT), it is possible to obtain the Industry 4.0 Maturity Level. Thus, 09 Maturity Levels were determined for this stage of the model, detailed in Chart 15:

Box. 15 - Industry 4.0 maturity levels

Level	Description
Leve	The company is at a critical level state regarding Industry 4.0. It needs to invest heavily in TT in order to start investing in the application of 4.0 concepts and technologies.

Level II	The company is in a state of alert regarding Industry 4.0. There is the existence of some components related to TT. However, there is no considerable existence of components related to Building Blocks 4.0. It is necessary to invest even more in TT, focusing on the search and implementation of BB4.0.
Level III	The company is in a state of alert in relation to Industry 4.0. TT related components are present, measured and controlled. However, there are no considerable components of Building Blocks 4.0. It is necessary to use the existing TT means in the company for a vision of I4.0 implementation. This will allow a better development of 4.0 concepts and technologies. This level indicates low influence of TT and I4.0.
Level IV	The company is in a state of alert in relation to Industry 4.0. There is the existence of some components related to Building Blocks 4.0. However, TT is at a critical level. It is necessary to invest strongly in TT, for a better development and application of the 4.0 concepts and technologies
Level V	The company is in an acceptable state regarding Industry 4.0. There is existence of both TT and BB4.0 components. It is still necessary to invest in Technology Transfer, for a better development and application of more 4.0 concepts and technologies. This level indicates strong influence of TT and I4.0.
Level VI	The company is in an acceptable state regarding Industry 4.0. There are Building Blocks 4.0 components, and even more TT components, which are measured and controlled. It is necessary to invest even more in TT and direct its focus to the 4.0 concepts, so Industry 4.0 will have a better performance.
Level VII	The company is in an acceptable state regarding Industry 4.0. The components related to Building Blocks 4.0 are present, measured and controlled. However, TT is at a critical level. Strong investment in TT is needed, for a better development and enhancement of 4.0 concepts and technologies This level indicates low TT and I4.0 influence.
Level VIII	The company is in an acceptable state regarding Industry 4.0. There are the TT components, and even more the Building Blocks 4.0 components, which are measured and controlled. It is still necessary to invest in Technology Transfer, for a better performance of the 4.0 concepts and technologies.
Level IX	The company is in an optimal state regarding Industry 4.0. There is a large existence of both TT and BB4.0 components, both are measured and controlled. At this stage, it is necessary to focus on Continuous Improvement to reach an optimal level of Industry 4.0. This level indicates strong influence of TT and I4.0.

Source: Own Authors (2021)

From levels I to IX, the production system becomes increasingly automated, flexible and intelligent. The components of both TT and BB4.0 are based on the low level, which means that this structure works in sequence. The ideal level of Industry 4.0 would be a level X - Continuous Improvement (intersection of levels 4 of TT and BB4.0). This level needs 100% implementation of the components of Technology Transfer and Building Blocks 4.0. Thus, it is necessary to have a structured plan of Continuous Improvement so that the maturity of Industry 4.0 is maintained at this level in the organisation.

The model structure was based on Qin and Liu's (2016) model, however, with a totally different context. The authors combine the levels of Intelligence and Engineering, to analyse the maturity of I4.0 technologies. In this work, it is the levels of TT and BB4.0 that are combined, to analyze the maturity of I4.0 in the company.

Thus, Figure 24 presents the I4.0-TT Maturity Model.

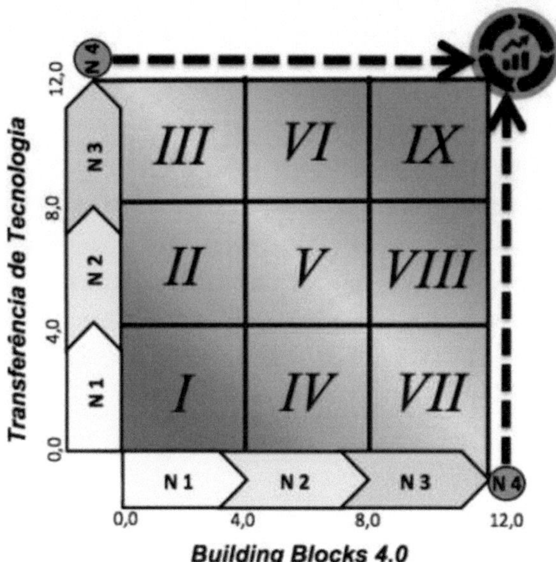

Figure 24 - I4.0-TT Maturity Model
Source: Author (2021)

With this model it is possible to analyse the maturity level of Industry 4.0 through Technology Transfer and *Building Blocks* 4.0. Thus, it is possible to verify if TT and I4.0 are actually correlated, and if they present a mutual influence.

With this model in place, it is also possible to visually analyse the Maturity Level of each Dimension.

7 APPLICATION OF THE TT-I4.0 MATURITY MODEL
7.1 APPLICATION OF THE QUESTIONNAIRES

Before collecting data, this study established the content validity of the survey through an extensive literature review, discussions with practitioners and academic experts in the field and a pre-test of the survey.

To validate the questionnaire and to qualify this research, a pilot test was applied at a company. This company, which we will call Company X, is a multinational located in the Campos Gerais region (Paraná - Brazil). Its branch is in packaging. This company has production control in its processes and products, as well as, aims at continuous improvement and implementation of new technologies related to the Fourth Industrial Revolution.

After the pilot test, the model was applied to companies from different sectors in the Campos Gerais region (Paraná - Brazil).

For Questionnaire 1, which aims to analyse the influence between TT and I4.0, the Sábato Triangle was used to select respondents at this stage. The Sábato Triangle is a model that presents the links between science, industry and government, which has informed science policy discussions across Latin America. It was developed in the 1960s and 1970s by Jorge Alberto Sábato.

Thus, 02 representatives from government institutions, 02 representatives from industry and 02 representatives from universities were randomly selected to answer this first questionnaire. All respondents have knowledge in Industry 4.0 and in Technology Transfer. Figure 25 presents the Sábato's Triangle used as a basis for choosing the respondents.

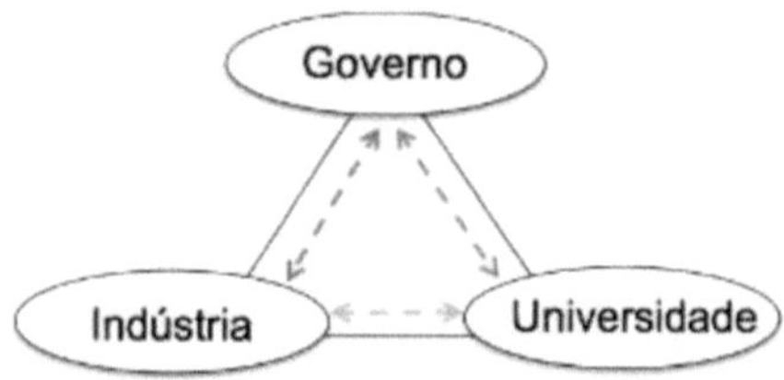

Figure 25 - Sábato's Triangle
Source: Adapted from Sábato and Botana (1975)

For Questionnaire 2, which aims to analyse the maturity of Industry 4.0 Concepts and Technologies and Technology Transfer, seven large companies were selected to answer the questions. All these companies intend to implement Industry 4.0 concepts and technologies. These companies were chosen for the reason that they have partnerships with the Federal Technological University of Paraná.

The seven industries where the model was applied are located in the state of Paraná. The choice was made to apply the model to companies in different fields so as to make the tool as generic as possible so as to subsequently replicate it in other productive fields and also obtain a vision of maturity in different sectors. For reliability reasons, industries will be named A, B, C, D, E, F and G. Chart 16 presents the industrial segment of each company.

Box. 16 - Industrial segment of the companies

COMPANY	INDUSTRIAL SEGMENT
A	Packaging
B	Food
C	Cosmetics
D	Automotive
E	Pulp and paper
F	Building materials
G	Autoparts

Source: Author (2021)

As for the application criteria, the questionnaire was applied to employees who have knowledge of all industry sectors and who are directly linked to the development process of Industry 4.0 in their respective industries.

The provision of the questionnaires was through the *Google Forms* instrument. After the questionnaires were created, the link for responses was made available by e-mail to the manager of each company analysed.

Data obtained was compiled into spreadsheets with the aid of *Excel software.* The corporate name of the company was not disclosed since fictitious names were determined so as to ensure anonymity during data tabulation. All managers received a reliability term signed by the researcher, who provided an individual report of his final diagnosis to each company that wished to, by way of knowledge.

7.2 MATHEMATICAL PROCESS

With the application of both questionnaires in the companies described, it was possible to develop the entire mathematical process presented in Step 5 of the methodology (Section 6.3.5).

Stage 5 (mathematical model) is divided into two parts, each with its own objectives. Thus, all the steps followed in the application of the model in each of these parts are presented below.

7. 2.1 Part One

With the application of Questionnaire 1, the 06 specialists previously described identified the level of influence between all Technology Transfer and Industry 4.0 dimensions, on a scale of 0 to 4. After this, the average of all responses was conducted and the following matrix (Table 1) was developed to thus begin the DEMATEL method.

Table 1 - Level of influence between all dimensions

Dimensions		**People**		**Process**		**Technologies**		**Strategy and Organisation**		**Integration**	
		HR	***HMI***	***PTT***	***PI4.0***	***DC***	***T4.0***	***CO***	***MN***	***MTT***	***IS***
People	***HR***	0.00	3.67	3.83	3.50	3.33	3.33	3.00	2.67	3.17	3.17
	HMI	3.17	0.00	2.67	2.67	2.17	2.50	2.33	1.33	2.00	3.00
Process	***PTT***	2.83	2.67	0.00	1.83	3.50	2.83	2.83	2.33	3.67	3.33
	PI4.0	3.17	3.17	2.50	0.00	2.00	3.67	3.00	2.67	2.50	3.17
Technologies	***DC***	3.17	3.17	4.00	3.50	0.00	2.83	2.67	2.50	3.00	2.83
	T4.0	3.17	3.50	2.67	3.50	1.83	0.00	2.67	3.17	2.50	3.50
Strategy and Organisation	***CO***	2.83	3.00	3.50	3.00	3.83	3.17	0.00	3.83	3.67	3.33
	MN	3.00	2.00	3.00	3.17	3.33	3.67	3.83	0.00	3.17	2.83
Integration	***MTT***	2.50	2.50	3.33	2.33	3.17	3.17	2.50	1.83	0.00	2.50
	IS	3.17	2.00	2.83	3.67	2.83	2.83	1.67	3.00	3.00	0.00

Source: Own authorship (2020)

From this initial matrix, the direct relationship matrix D was built using Equations 1 and 2. After the matrix D, the total relationship matrix T was built according to Equation 3. And thus, it was possible to calculate the sum of each row □□ of matrix T, which represents the total impact that each element has on the set of elements studied, and the sum of each column □□ of matrix T, which represents the total impact received by each element on the set of elements studied.

For each dimension, it was also calculated the sum □□ + □□, which represents the total involvement that each element has in the set of elements studied and the difference □□ - □□, which represents the net effect that each element has in the set of elements studied.

Table 2 presents the results found in the first step of the DEMATEL method: the sum of each row □□, of each column □□, the sum □□ + □□ and the difference □□ - □□.

Table 2 - Results found using the DEMATEL method

Dimensions	*laugh*	*ci*	*ri + ci*	*ri - ci*
Human Resources	0.1107	0.1021	0.2128	0.0086
Human-Machine Interface	0.0840	0.0978	0.1818	-0.0137
TT Cases	0.0979	0.1064	0.2043	-0.0086
Processes I4.0	0.0980	0.1025	0.2005	-0.0045
Dissemination of Knowledge	0.1040	0.0982	0.2023	0.0058
Technologies 4.0	0.1000	0.1053	0.2053	-0.0053
Organisational Culture	0.1129	0.0930	0.2059	0.0199
Business Models	0.1064	0.0893	0.1956	0.0171
TT Modes	0.0910	0.1008	0.1918	-0.0098
Systems Integration	0.0952	0.1045	0.1997	-0.0093

Source: Own Authors (2021)

7.2.2 Part Two

In the previous part it was possible to obtain the weights (*w*) for each of the dimensions of this study, so as to begin the second part of this stage of the model. These weights were obtained from the sum of each row □□, which represents the overall impact given by each factor.

Seven managers answered Questionnaire 2, which aims to analyse the maturity of Concepts and Technologies 4.0 and Technology Transfer in the company, giving grades from 0 to 4 for each of the questions.

After collecting the answers to the questionnaires applied, and obtaining the weights (*w*) of the dimensions, it was possible to conclude the mathematical development of the model proposed in this research. As of Equations 6 and 7 the final Value (V_d) and the Maturity of each Dimension (M_d) were extracted. Thus, with the sum of the dimension values (V_d), it was obtained the Maturity of each Factor (M_f) for each one of the companies, by using Equation 8.

Thus, Table 3 presents the values found for each of the dimensions and factors.

Table 3 - Values found for each factor and dimension

COMPANY	Factors			Technology Transfer					Building Blocks 4.0		
	Dimensions	HR	PTT	DC	CO	MTT	HMI	PI4.0	T4.0	MN	IS
	w	*0.111*	*0.098*	*0.104*	*0.113*	*0.091*	*0.084*	*0.098*	*0.100*	*0.106*	*0.095*
A	**Value Dimension (*Vd*)**	1.439	0.979	1.352	1.580	1.911	0.672	1.470	1.399	1.489	0.571
	Maturity Dimension (*Md*)	0.288	0.244	0.270	0.225	0.212	0.224	0.183	0.155	0.297	0.114
	Maturity Factor (*Mf*)			**7.2607**				**5.6021**			
B	**Value Dimension (*Vd*)**	0.221	0.783	0.312	0.790	0.273	0.420	0.000	0.199	0.000	0.190
	Maturity Dimension (*Md*)	0.044	0.195	0.062	0.112	0.030	0.140	0.000	0.022	0.000	0.038
	Maturity Factor (*Mf*)			**2.3795**				**0.8105**			
C	**Value Dimension (*Vd*)**	1.217	0.978	0.728	2.257	1.183	0.756	1.960	1.099	0.425	0.666
	Maturity Dimension (*Md*)	0.243	0.245	0.145	0.322	0.131	0.252	0.245	0.122	0.085	0.133
	Maturity Factor (*Mf*)			**6.3646**				**4.9079**			
D	**Value Dimension (*Vd*)**	0.664	1.174	0.936	1.692	2.092	0.504	0.980	1.999	0.745	0.952
	Maturity Dimension (*Md*)	0.132	0.293	0.187	0.242	0.233	0.168	0.123	0.222	0.149	0.190
	Maturity Factor (*Mf*)			**6.5605**				**5.1801**			
E	**Value Dimension (*Vd*)**	1.218	1.077	1.040	2.144	1.820	0.588	1.470	1.799	1.489	0.952
	Maturity Dimension (*Md*)	0.244	0.269	0.208	0.306	0.202	0.196	0.184	0.200	0.298	0.190
	Maturity Factor (*Mf*)			**7.2985**				**6.2987**			
F	**Value Dimension (*Vd*)**	0.553	0.783	0.312	0.564	1.183	0.420	0.392	0.000	0.213	0.286
	Maturity Dimension (*Md*)	0.111	0.196	0.062	0.081	0.131	0.140	0.049	0.000	0.043	0.057
	Maturity Factor (*Mf*)			**3.3957**				**1.3105**			
G	**Value Dimension (*Vd*)**	1.107	1.077	0.624	1.242	1.092	0.504	1.078	0.900	0.425	0.381
	Maturity Dimension (*Md*)	0.221	0.269	0.125	0.177	0.121	0.168	0.135	0.100	0.085	0.076
	Maturity Factor (*Mf*)			**5.1410**				**3.2882**			

Source: Own Authors (2021)

This closes the mathematical procedure of the model. Once this is done, it is possible to analyse the results of each company using the model proposed in this study. All results, analyses and discussions will be presented in the next chapter.

8 RESULTS AND DISCUSSIONS

8.1 INFLUENCE OF THE DIMENSIONS

After applying the proposed model, in the first part of the model it was possible to obtain the results referring to the dimensions that influence each other, and also the weights of the dimensions.

With the DEMATEL approach, it was possible to classify the components studied into 2 groups: the influenced, who have negative *r-c* and the influencers, who have positive *r-c.* This happens because the influencers exercise a degree of impact greater than the degree of impact they receive within the set studied. And those influenced receive a degree of impact greater than the degree of impact they exert on the other components within the studied set.

Figure 26 presents a graph which represents this situation found in the model. Dimensions which are above the x-axis (*ri* + *ci*) are the dimensions responsible for influencing the others and those which are below this axis are the dimensions which are influenced. In this graph, one also analyses that the further to the right of the same, that is, the higher the value of *ri* + *ci, the* greater the total involvement each dimension has in this set under study.

Figure 26 - Influencing and influenced dimensions, and total involvement that each dimension has in the whole
Source: Own Authors (2021)

One notices that the dimension that has the greatest total involvement in the whole is *Human Resources.* This dimension influences the others so that many things happen, it is the "lung" of this system, it is a fundamental dimension for the entire company. For this reason, it is the dimension that has the greatest total involvement.

To facilitate visualisation, Table 4 was prepared with the dimensions which are responsible for influencing the others and those which are influenced.

Table 4 - Influenced and influencing dimensions

Influencing Dimensions	Influenced Dimensions
Organisational Culture	TT Modes
Business Models	Human-Machine Interface
Human Resources	Process I4.0
Dissemination of Knowledge	Technologies 4.0
	TT Process
	Systems Integration

Source: Own Authors (2021)

Three of the influencing dimensions refer to Technology Transfer (Organisational Culture, Human Resources and Knowledge Dissemination) and only one refers to Building Blocks 4.0 (Business Models). This proves the fact that TT is a strategic factor to leverage the concepts and technologies 4.0 in companies.

In Figure 27 it is possible to perceive the main correlation pairs; that is, the most important and relevant influence directions of this set, presented by means of the total relation matrix DEMATEL T (above 0.70). The direction of the arrow is from influencing dimension to influenced dimension, and the number above is its degree of relevance. Thus, it is possible to visualise which dimensions influence each one of them, and their intensity.

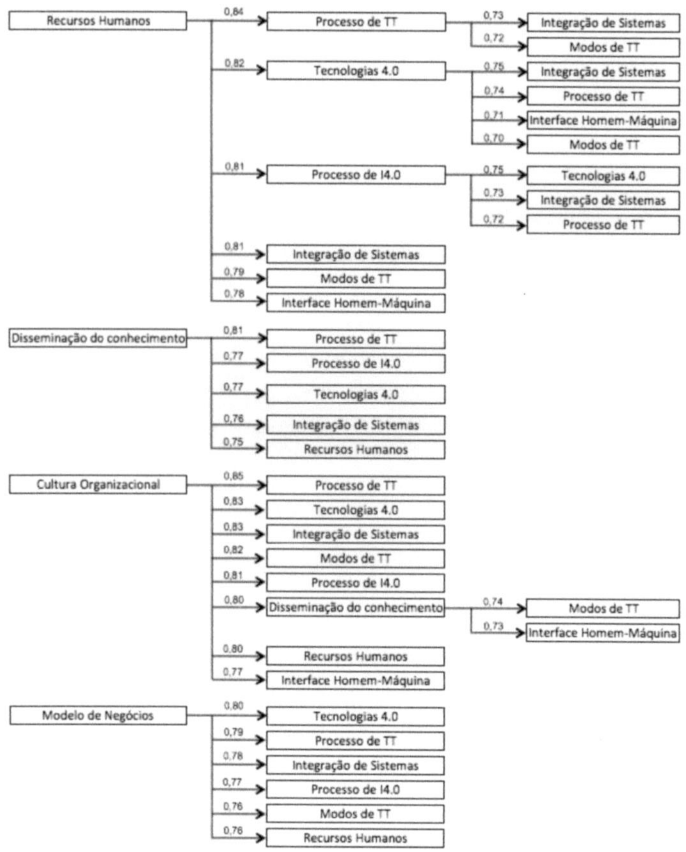

Figure 27 - Most relevant influences of the set
Source: Own Authors (2021)

The objective of knowing which are the influencing and influenced dimensions is to be able to design strategies for companies to raise their I4.0 maturity level. Furthermore, this model is based on the influences between the dimensions.

8.2 MATURITY LEVEL I4.0-TT

With the results obtained in the second part of the model, it is possible to identify the Industry 4.0 Maturity Level of each company studied. Furthermore, it is also possible to identify the level of Maturity of each Dimension so that a diagnosis of each company may be presented.

Figure 28 presents the placement of Companies A, B, C, D, E, F and G in the I4.0-TT Maturity Model.

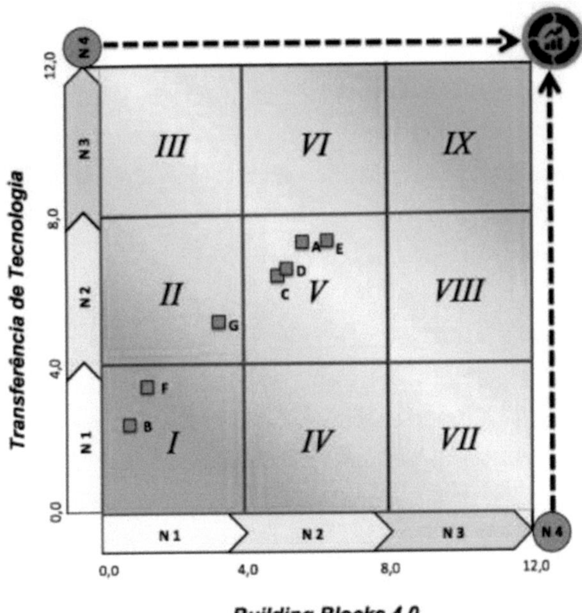

Figure 28 - Maturity Level I4.0-TT for the companies studied
Source: Own Authors (2021)

Companies A, C, D and E are at Level V of Industry 4.0. This means that they are in an acceptable state. There is the existence of both TT and BB4.0 components. It is still necessary to invest in Technology Transfer, for a better performance and application of more 4.0 concepts and technologies. This level indicates a strong influence of TT and I4.0.

Company G is at Level II of Industry 4.0. It is an alert state in relation to Industry 4.0. There is the existence of some components related to TT. However, there is no considerable existence of components related to *Building Blocks* 4.0. It is necessary to invest even more in TT, having as focus the search and implementation of BB4.0.

On the other hand, Companies B and F are at Level I of Industry 4.0. Both companies present a critical level status regarding Industry 4.0. It is necessary to invest heavily in TT so as to start investing in the application of 4.0 concepts and technologies.

For each company analysed, a report was made on how the company stands both for Technology Transfer and for Industry 4.0, and also, strategies that could be used to raise the company's maturity level. This diagnosis is presented below in Section 8.3.

With this model, besides verifying the maturity level of Industry 4.0 that the organisation is, it is possible to analyse if there is really influence between TT and I4.0. If a lot of dispersion of results occurred (mainly level III and VII), it meant that this influence was not so strong. As the results concentrated mainly in levels I, V and IX, it means that there is really this influence between TT and I4.0, proposed in the theoretical model.

8.3 COMPANY DIAGNOSIS

With the influence of the dimensions and the level of maturity of I4.0 of each company, it is possible to perform an individual diagnosis of each participating company, and present strategies and actions needed to evolve to the next level.

Industry 4.0 is a very complex and difficult topic to implement. For this reason, strategies will be presented so that through the proposed maturity model, the company may advance in this context.

Firstly, a path has been outlined in the model for the company to aim for. It is called the ideal path to achieve continuous improvement in Industry 4.0 (Figure 29).

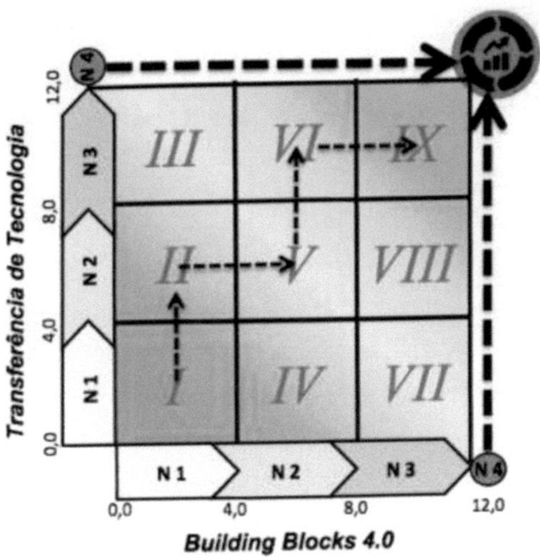

Figure 29 - Ideal path to achieve continuous improvement in I4.0
Source: Own Authors (2021)

This path aims to achieve first a higher maturity level of Technology Transfer, to subsequently reach a higher maturity level of Building Blocks 4.0. With this, a higher maturity level of Industry 4.0 will be reached. This order is due to the fact that three of the four influencing dimensions are related to TT, that is, TT becomes an ally to BB 4.0 for the evolutionary implementation of Industry 4.0 in organisations.

Focusing on the I4.0 maturity level to be reached, the next step is to identify the current stage of each of the TT and BB4.0 dimensions. To reach the next level in each dimension, many actions are required. To facilitate this process, the suggested strategy is to list priorities in the action plan, in the following order:

1) Actions related to dimensions that influence the dimensions most lacking in the company;

2) Actions related to the influencing dimensions (Human Resources, Dissemination of Knowledge, Organizational Culture and Business Model);

3) Actions related to the dimensions, taking into account feasibility and necessary investment.

Given the above, the individual diagnoses of each of the companies studied will be presented below, together with suggestions for actions and strategies so that the maturity level may be raised.

The suggestions presented are directly related to the shortages identified through the answers obtained in the questionnaires. However, companies must evaluate if the advances in the levels are compatible with their respective objectives.

8.3.1 Company A

According to the model presented (Figure 33), Company A is situated at Level V (Acceptable) of Industry 4.0. This means that there is existence of both the TT and BB4.0 components, but there is still much to improve.

Figure 30 presents the levels in which each dimension of TT and BB 4.0 fits, from the results obtained.

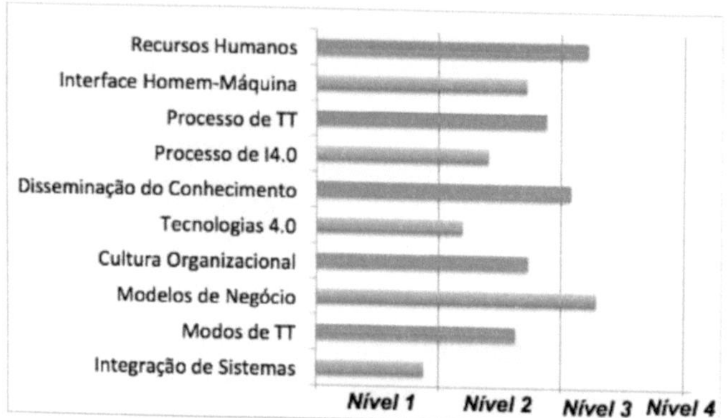

Figure 30 - Levels of maturity of the dimensions in Company A
Source: Own Authors (2021)

One notices that three of these dimensions are at Level 3: Human Resources, Knowledge Dissemination and Business Models. Of these, two are related to Technology Transfer (HR and KD) and one to Building Blocks 4.0 (KD). The dimension that presented the lowest level was Systems Integration, a BB4.0 dimension.

In order to find measures to raise the company's Industry 4.0 maturity level, Table 17 presents the current level of each dimension, and actions needed for it to reach the next level.

Box. 17 - Suggested actions for each dimension of Company A

COMPANY	DIMENSION	CURRENT LEVEL	NEXT LEVEL	ACTIONS

	Human Resources	Level 3	Level 4	Provide training in technical and managerial skills related to I4.0 in more than half of the sectors.
	Human-Machine Interface	Level 2	Level 3	Encourage autonomy and freedom of creation.
	TT Process	Level 2	Level 3	Enter into partnerships with TT offices, development agencies, among other TT facilitators; Transfer technologies more frequently to other companies.
	I4.0 Process	Level 2	Level 3	Implement in more than half of the sectors: flexibility in producing highly customised products; integrated sensor system; intelligent maintenance; digital modeling capabilities and performance simulation.
	Dissemination of Knowledge	Level 3	Level 4	Investing in research, with the aim of having patents and scientific publications.
A	Technologies 4.0	Level 2	Level 3	Implement in more than half of the sectors: embedded and intelligent systems; Big Data; Track and Trace; Cloud Manufacturing; Augmented Reality; Additive Manufacturing; Artificial Intelligence.
	Organisational Culture	Level 2	Level 3	Implement more consultancies in the company; Participate in networking events; Set up partnerships with masters and doctorate programmes; Share the results of employees at seminars.
	Business Models	Level 3	Level 4	Implement performance indicators for actions related to I4.0
	TT Modes	Level 2	Level 3	Develop strong partnerships with Universities.
	Systems Integration	Level 1	Level 2	Implement real time information exchange systems with your supply chain, and bi-directional updating of information between the facilities and real equipment with digital copies.

Source: Author (2021)

However, there are many actions that are necessary for the Industry 4.0 Maturity level to be raised in the company. To this end, the strategy to be used will be to prioritise the dimensions that influence the others.

As the *Systems Integration* dimension is the most precarious at Company A, it was observed in Figure x that four dimensions (Human Resources, Knowledge Dissemination, Organizational Culture and Business Model) influence it.

Thus, it is suggested that these four dimensions need to be prioritised in the action plans. Consequently, *Systems Integration* and other dimensions will benefit from the improvements achieved in these four dimensions.

Subsequently, the feasibility and investment required for each of the actions presented must be analysed, in order to list which actions should be carried out.

For Company A to advance to Level VI, it must invest heavily in Technology Transfer. This will enable better performance and the application of more 4.0 concepts and technologies.

8.3.2 Company B

According to the model presented (Figure 33), Company B is located at Level I of Industry 4.0. It is considered a critical level, where both 4.0 concepts and technologies, as well as Technology Transfer, are practically non-existent. There is still much to advance in this company.

Figure 31 presents the levels into which each dimension of TT and BB 4.0 fits, from the results obtained.

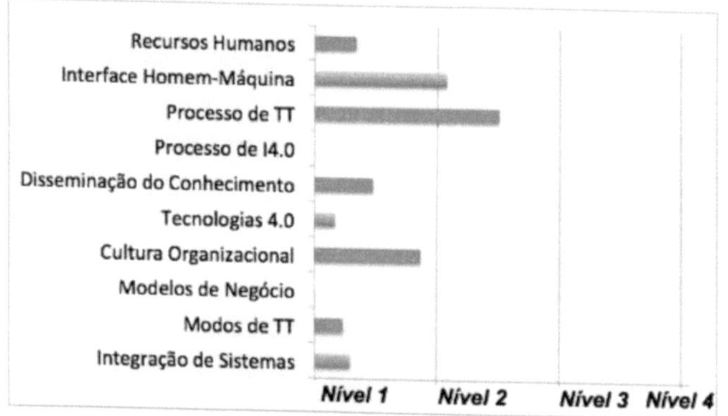

Figure 31 - Levels of maturity of the dimensions in Company B
Source: Own Authors (2021)

It can be noticed that only two of these dimensions are at Level 2: *Human-Machine Interface* and *TT Process*. The remaining dimensions are all at Level 1, and two are still at Level 0 (*I4.0 Process* and *Business Models*).

In order to find measures to raise the company's Industry 4.0 maturity level, Table 18 presents the current level of each dimension, and actions needed for it to reach the next level.

Box. 18 - Suggested actions for each dimension of Company B

COMPANY	DIMENSION	CURRENT LEVEL	NEXT LEVEL	ACTIONS
B	Human Resources	Level 1	Level 2	Implementation of

				more comprehensive training in the company for all employees; Start a training of the skills and competences of I4.0; Implement interdisciplinarity in the work groups.
	Human-Machine Interface	Level 2	Level 3	Implement incentive programmes for new ideas and creations, as well as autonomy; Invest heavily in training in new technologies.
	TT Process	Level 2	Level 3	Analyse the feasibility of partnerships with TT offices, development agencies, among other TT facilitators.
	I4.0 Process	Level 0	Level 1	Start the implementation of: customised products, automation of some processes, integrated sensor system, intelligent maintenance, modelling and simulation.
	Dissemination of Knowledge	Level 1	Level 2	To start an investment in research, with the aim of having patents and scientific publications; To start digitalising and sharing business processes internally and with other partners.
	Technologies 4.0	Level 1	Level 2	Start implementing: products with embedded and intelligent systems; traceability of the final product; Big Data; Cloud Manufacturing; Information Security;

				Augmented Reality; Additive Manufacturing; Autonomous Robots.
	Organisational Culture	Level 1	Level 2	Start partnerships with master's and doctoral programmes; Increase the number of internships in the organisation; Participate in networking events; Start programmes to encourage innovation.
	Business Models	Level 0	Level 1	Initiate the implementation of: Central coordination for I4.0 related actions; Performance indicators for I4.0; Planning and investment for 4.0 technologies; Marketing 4.0.
	TT Modes	Level 1	Level 2	Start strong partnerships with Universities; Start sharing knowledge with customers, suppliers, and among the different sectors of the company.
	Systems Integration	Level 1	Level 2	Start implementing: product lifecycle management; bi-directional updating of information between real installations and equipment with digital copies; real-time information exchange.

Source: Author (2021)

However, there are many actions that are necessary for the Industry 4.0 Maturity level to rise in the company. To this end, the strategy to be used will be to prioritise the dimensions that influence the others.

The dimensions *TT Process* and *Business Models* are the ones that have a lower level of maturity in Company B. Thus, it was observed in Figure x that four dimensions (Human Resources, Knowledge Dissemination, Organisational Culture and Business Models) influence TT Process; and the *Business Models* dimension influences several other dimensions (it is considered an influencing dimension).

The fact that the *Business Models* dimension is at Level 0 and is influential, is interfering negatively in all other dimensions. For this reason the company is at such a low level of maturity of I4.0

Thus, it is suggested that the *Business Models* dimension should be prioritised in the action plans, followed by *Human Resources, Knowledge Dissemination* and *Organisational Culture*. Consequently, the *TT process* and other dimensions will benefit from the improvements achieved in these four dimensions mentioned.

Subsequently, the viability and investment required for each of the actions presented must be analysed, in order to list which actions should be carried out.

For Company B to advance to Level II, it must invest heavily in Technology Transfer. Thus, to thus start investing in the application of the 4.0 concepts and technologies.

8.3.3 Company C

According to the model presented (Figure 33), Company C is situated at Level V (Acceptable) of Industry 4.0. This means that there is existence of both the TT and BB4.0 components, but there is still much to improve.

Figure 32 presents the levels in which each dimension of TT and BB 4.0 fits, from the results obtained.

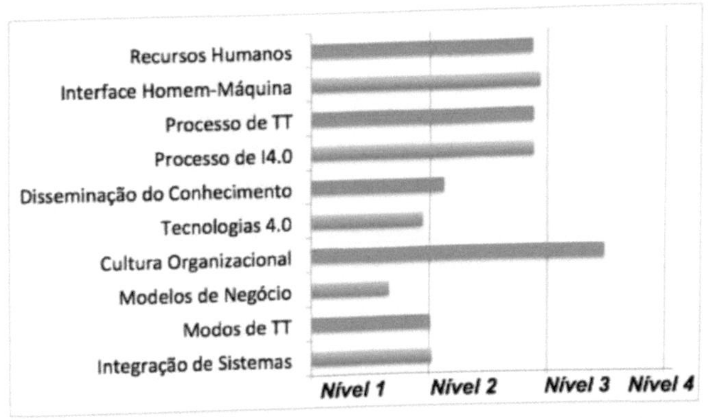

Figure 32 - Levels of maturity of the dimensions in Company C
Source: Own Authors (2021)

One notices that only one of these dimensions is at Level 3: *Organizational Culture*. The dimensions that presented a lower level were *Technologies 4.0* and *Business Models,* both BB 4.0 dimensions. The other dimensions are all at Level 2.

In order to find measures to raise the company's Industry 4.0 maturity level, Table 19 presents the current level of each dimension, and actions needed for it to reach the next level.

Frame. 19 - Suggested actions for each dimension of Company C

COMPANY	DIMENSION	CURRENT LEVEL	NEXT LEVEL	ACTIONS
C	Human Resources	Level 2	Level 3	Carry out I4.0 related technical and managerial skills training in more than half of the sectors; Implement an interdisciplinary approach in all working groups.
	Human-Machine Interface	Level 2	Level 3	All collaborators must have: autonomy and freedom for creation; master the technologies they use; be open to new technologies.
	TT Process	Level 2	Level 3	Enter into partnerships with TT offices, development agencies, among other TT facilitators; Transfer technologies more frequently to other companies.
	I4.0 Process	Level 2	Level 3	Implement in more than half of the sectors: intelligent maintenance; performance modelling and simulation; flexibility in producing highly customised products; integrated sensor system; adaptable production lines.
	Dissemination of Knowledge	Level 2	Level 3	To invest in research in order to have patents and

				scientific publications.
	Technologies 4.0	Level 1	Level 2	Initiate the implementation of: Embedded and intelligent systems; Augmented reality; Additive manufacturing; Autonomous robots; Artificial intelligence.
	Organisational Culture	Level 3	Level 4	Partner with master's and doctoral programmes; Participate in networking events; Initiate programmes to encourage innovation.
	Business Models	Level 1	Level 2	Implement performance indicators for actions related to I4.0; Implement Marketing 4.0.
	TT Modes	Level 2	Level 3	Initiate partnerships with Universities; Transfer technological knowledge from the R&D phase to the commercialisation of the product in the company, and between company projects.
	Systems Integration	Level 2	Level 3	Increase the level of connectivity of the company. Implement product life cycle management systems, and real-time information exchange with the supply chain.

Source: Author (2021)

However, there are many actions that are necessary for the Industry 4.0 Maturity level to rise in the company. To this end, the strategy to be used will be to prioritise the dimensions that influence the others.

The dimensions *Technologies 4.0* and *Business Models are* the most precarious at Company C. Thus, it was observed in Figure x that four dimensions (Human

Resources, Knowledge Dissemination, Organisational Culture and Business Models) influence *Technologies 4.0;* and the *Business Models* dimension influences several other dimensions (it is considered an influencing dimension).

Thus, it is suggested that it is necessary to prioritise these four dimensions in the action plans, starting with *Business Models*. Consequently, *Technologies 4.0* and other dimensions will benefit from the improvements achieved in these four dimensions mentioned above.

Subsequently, the viability and investment required for each of the actions presented must be analysed, in order to list which actions should be carried out.

For Company C to advance to Level VI, it must invest heavily in Technology Transfer. This will enable better performance and the application of more 4.0 concepts and technologies.

8.3.4 Company D

According to the model presented (Figure 33), Company D is situated at Level V (Acceptable) of Industry 4.0. This means that there is existence of both the TT and BB4.0 components, but there is still much to improve.

Figure 33 presents the levels in which each dimension of TT and BB 4.0 fits, from the results obtained.

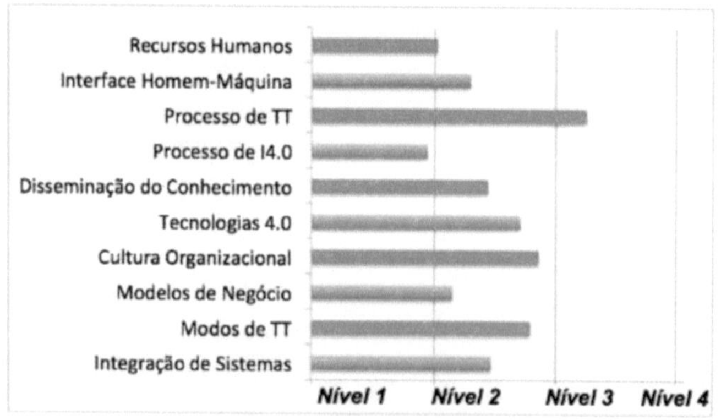

Figure 33 - Levels of maturity of the dimensions in Company D
Source: Own Authors (2021)

It can be seen that only one of these dimensions is at Level 3: *TT Process*. The dimension that presented the lowest level was *Process of I4.0.* The remaining dimensions are all at Level 2.

In order to find measures to raise the company's Industry 4.0 maturity level, Table 20 presents the current level of each dimension, and actions needed for it to reach the next level.

Table. 20 - Suggested actions for each dimension of Company D

COMPANY	DIMENSION	CURRENT LEVEL	NEXT LEVEL	ACTIONS
D	Human Resources	Level 2	Level 3	Implementation of more comprehensive training in the company for all employees; Start a training of the skills and competences of I4.0.
	Human-Machine Interface	Level 2	Level 3	Implement programmes to encourage new ideas and creations, as well as autonomy.
	TT Process	Level 3	Level 4	Partner with TT offices, development agencies, among other TT facilitators.
	I4.0 Process	Level 1	Level 2	Increase the automation of processes; Start implementing: intelligent maintenance, and digital modelling and simulation resources.
	Dissemination of Knowledge	Level 2	Level 3	To carry out investment in research, with the aim of having patents and scientific publications; To implement the digitalisation and sharing of business processes internally in the company.
	Technologies 4.0	Level 2	Level 3	Start the implementation of augmented reality,

				and continue the implementation and expansion of the other technologies already existing in the company.
	Organisational Culture	Level 2	Level 3	Initiate partnerships with master's and doctoral programmes.
	Business Models	Level 2	Level 3	Implement performance indicators for actions related to I4.0
	TT Modes	Level 2	Level 3	Initiate models for sharing relevant information with other companies in the supply chain for agile decision making
	Systems Integration	Level 2	Level 3	Implement product life cycle management systems.

Source: Author (2021)

However, there are many actions that are necessary for the Industry 4.0 Maturity level to rise in the company. To this end, the strategy to be used will be to prioritise the dimensions that influence the others.

The dimension *I4.0 Processes* is the one that presents the lowest level of maturity at Company D. Thus, it was observed in Figure x that four dimensions (Human Resources, Dissemination of Knowledge, Organizational Culture and Business Models) influence *I4.0 Processes.*

Thus, it is suggested that it is necessary to prioritise these four dimensions in the action plans. As a consequence, *4.0 Processes* and other dimensions will benefit from the improvements achieved in these four dimensions mentioned.

Subsequently, the feasibility and investment required for each of the actions presented must be analysed, in order to list which actions should be carried out.

For Company D to advance to Level VI, it must invest heavily in Technology Transfer. This will enable better performance and the application of more 4.0 concepts and technologies.

8.3.5 Company E

According to the model presented (Figure 33), Company E is situated at Level V (Acceptable) of Industry 4.0. This means that there is existence of both the TT and BB4.0 components, but there is still much to improve.

Figure 34 presents the levels in which each dimension of TT and BB 4.0 fits, from the results obtained.

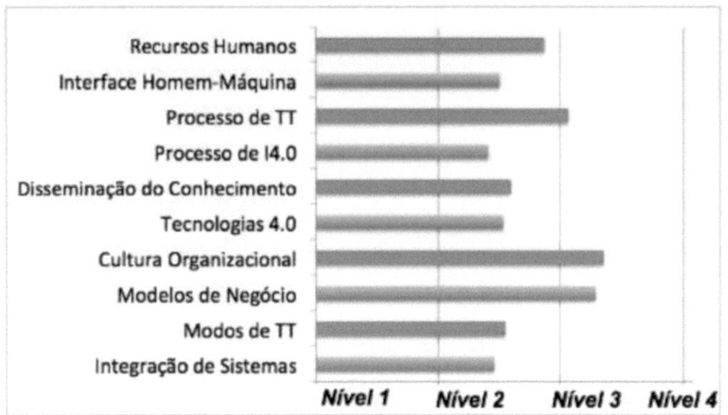

Figure 34 - Levels of maturity of the dimensions in Company E
Source: Own Authors (2021)

This was the company that presented a better performance than the others. It is noticed that all dimensions are between levels 2 and 3, being that the *TT Process, Organisational Culture* and *Business Models* are the ones with a higher maturity level.

It is observed that *Organizational Culture* and *Business Models* are influencing dimensions. One of the reasons that the company is advancing in the maturity of I4.0, is the fact that these dimensions are with a high level of maturity.

In order to find measures to raise the company's Industry 4.0 maturity level, Table 21 presents the current level of each dimension, and actions needed for it to reach the next level.

Box. 21 - Suggested actions for each dimension of Company E

COMPANY	DIMENSION	CURRENT LEVEL	NEXT LEVEL	ACTIONS
E	Human Resources	Level 2	Level 3	Provide training in technical and managerial skills related to I4.0 in more than half of the sectors.
	Human-Machine Interface	Level 2	Level 3	Implement incentive programmes for new ideas and creations for all employees, as well as autonomy.
	TT Process	Level 3	Level 4	Transfer technologies to other companies more frequently.
	I4.0 Process	Level 2	Level 3	Increase process automation; Implement in more sectors of the company: integrated sensor system, carry out integrated information exchanges in real time, adaptable production lines, intelligent maintenance, digital modelling and simulation resources.
	Dissemination of Knowledge	Level 2	Level 3	To invest in research in order to have patents and scientific publications.
	Technologies 4.0	Level 2	Level 3	Implement in some sectors of the company: production of products with embedded and intelligent systems, augmented reality, additive manufacturing, and artificial intelligence.
	Organisational Culture	Level 3	Level 4	Participate in more networking events; Make stronger partnerships with masters and doctoral programmes; Invest in the

				development of new ideas and innovations.
	Business Models	Level 3	Level 4	Invest in marketing 4.0; Conduct agile and digitised communication with customers for sales, service and information acquisition.
	TT Modes	Level 2	Level 3	Initiate company-wide information sharing models, across the supply chain, with customers and suppliers.
	Systems Integration	Level 2	Level 3	Exchange information in real time with your supply chain.

Source: Author (2021)

For the company to continue advancing in Industry 4.0, it is suggested that the four influencing dimensions be prioritised in the action plan: Human Resources, Knowledge Dissemination, Organisational Culture and Business Models.

In this way, other dimensions will benefit from the improvements obtained in these four dimensions mentioned.

Subsequently, the feasibility and investment required for each of the actions presented must be analysed, in order to list which actions should be carried out.

For Company E to advance to Level VI, further investment in Technology Transfer is required. This will enable better performance and the application of more 4.0 concepts and technologies.

8.3.6 Company F

According to the model presented (Figure 33), Company F is located at Level I of Industry 4.0. It is considered a critical level, where both 4.0 concepts and technologies, as well as Technology Transfer, are practically non-existent. There is still much to advance in this company.

Figure 35 presents the levels into which each dimension of TT and BB 4.0 fits, from the results obtained.

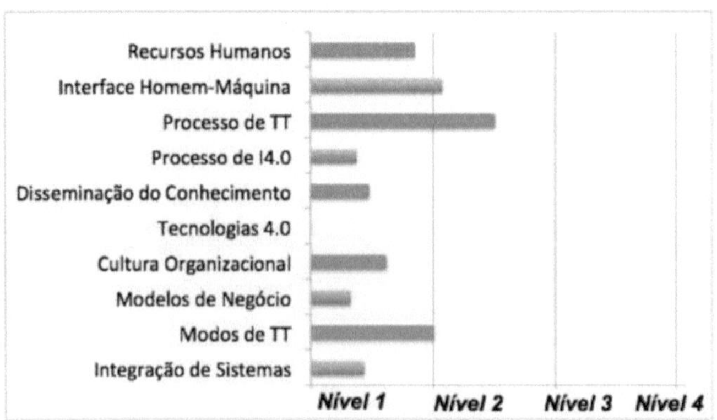

Figure 35 - Levels of maturity of the dimensions in Company F
Source: Own Authors (2021)

It can be seen that only three of these dimensions are at Level 2: *Human-Machine Interface*, *TT Process* and *TT Modes*. The remaining dimensions are all at Level 1, and one is still at Level 0 (*I4.0 Technologies*).

In order to find measures to raise the company's Industry 4.0 maturity level, Table 22 presents the current level of each dimension, and actions needed for it to reach the next level.

Box. 22 - Suggested actions for each dimension of Company F

COMPANY	DIMENSION	CURRENT LEVEL	NEXT LEVEL	ACTIONS
F	Human Resources	Level 1	Level 2	Implementation of more comprehensive training in the company for all employees; Start a training of the skills and competences of I4.0; Implement interdisciplinarity in the work groups.
	Human-Machine Interface	Level 2	Level 3	Implement programmes to encourage new ideas and creations, as well as autonomy.
	TT Process	Level 2	Level 3	Partner with TT offices, development agencies, among other TT facilitators.
	I4.0 Process	Level 1	Level 2	Start implementing: automation in processes, integrated sensor system, intelligent maintenance, digital modeling and simulation resources.
	Dissemination of Knowledge	Level 1	Level 2	To start an investment in research, in order to have patents and scientific publications; To start the digitalisation and sharing of business processes internally and with other partners.
	Technologies 4.0	Level 0	Level 1	Start implementing: products with embedded and intelligent systems, end-product traceability, Big Data, cloud computing, information security, augmented reality, additive manufacturing, autonomous robots.

	Organisational Culture	Level 1	Level 2	Initiate partnerships with master's and doctoral programmes; Share collaborators' results in seminars.
	Business Models	Level 1	Level 2	Initiate the implementation of: Central coordination for I4.0 related actions; Performance indicators for I4.0; Planning and investment for 4.0 technologies; Marketing 4.0.
	TT Modes	Level 1	Level 2	Carry out strong partnerships with universities; Initiate information sharing models across the company, across the supply chain, with customers and suppliers.
	Systems Integration	Level 1	Level 2	Implement real time information exchange systems with your supply chain, and bi-directional updating of information between the facilities and real equipment with digital copies.

Source: Author (2021)

However, there are many actions that are necessary for the Industry 4.0 Maturity level to rise in the company. To this end, the strategy to be used will be to prioritise the dimensions that influence the others.

The Technologies 4.0 dimension is the one with the lowest maturity level at Company F. Therefore, it was observed in Figure x that five dimensions (Human Resources, Knowledge Dissemination, Organisational Culture and Business Models and I4.0 Process) influence *Technologies 4.0.*

Thus, it is suggested that it is necessary to prioritise these five dimensions in the company's action plans. Consequently, *Technologies 4.0* and other dimensions will benefit from the improvements obtained in these dimensions mentioned above.

Subsequently, the feasibility and investment required for each of the actions presented must be analysed, in order to list which actions should be carried out.
For Company F to advance to Level II, it must invest heavily in Technology Transfer. Thus, to thus start investing in the application of the 4.0 concepts and technologies.

8.3.7 Company G

According to the model presented (Figure 33), Company G is located at Level II of Industry 4.0. It is considered an alert state in relation to Industry 4.0. There is the existence of some components related to TT. However, there is no considerable existence of components related to Building Blocks 4.0.
Figure 36 presents the levels in which each dimension of TT and BB 4.0 fits, from the results obtained.

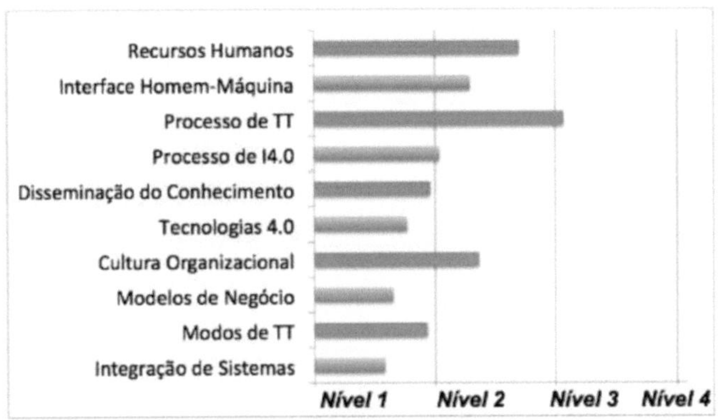

Figure 36 - Levels of maturity of the dimensions in Company G
Source: Own Authors (2021)

It can be seen that only one of these dimensions is at Level 3: *TT Process*. The dimensions that presented a lower level were: *Technologies 4.0*, *Business Models* and *Systems Integration*, all dimensions of BB4. The other dimensions are at Level 2 or very close to reaching this level.

This is the company that most differentiated itself from the others, it was the only one that presented a Level of Technology Transfer (Level 2) different from *Building Blocks* 4.0 (Level 1). This means that the company is in transition to Level V of

Industry 4.0. The goal now should be to raise the level of BB4.0 from Level 1 to Level 2.

In order to find measures to raise the company's Industry 4.0 maturity level, Table 23 presents the current level of each dimension, and actions needed for it to reach the next level.

Box. 23 - Suggested actions for each dimension of Company G

COMPANY	DIMENSION	CURRENT LEVEL	NEXT LEVEL	ACTIONS
G	Human Resources	Level 2	Level 3	Provide training in technical and managerial skills related to I4.0 in more than half of the sectors.
	Human-Machine Interface	Level 2	Level 3	Implement programmes to encourage new ideas and creations, as well as autonomy.
	TT Process	Level 3	Level 4	Partner with TT offices, development agencies, among other TT facilitators.
	I4.0 Process	Level 2	Level 3	Carry out implementation of: integrated sensor system, and intelligent maintenance.
	Dissemination of Knowledge	Level 1	Level 2	To start an investment in research, in order to have patents and scientific publications; To start the digitalisation and sharing of business processes internally and with other partners.
	Technologies 4.0	Level 1	Level 2	Start implementing: augmented reality, end-product traceability, additive manufacturing.
	Organisational Culture	Level 2	Level 3	Invest in consultancies for the company; Partner with masters and doctorate

				programmes.
	Business Models	Level 1	Level 2	Initiate the implementation of: Central coordination for I4.0 related actions; Performance indicators for I4.0; Planning and investment for 4.0 technologies; Marketing 4.0.
	TT Modes	Level 1	Level 2	Carry out strong partnerships with universities; Initiate information sharing models across the company, between the supply chain, with customers and suppliers.
	Systems Integration	Level 1	Level 2	Implement real time information exchange systems with your supply chain, and bi-directional updating of information between the facilities and real equipment with digital copies.

Source: Author (2021)

However, many are the necessary actions so that the Industry 4.0 Maturity level is raised in the company. To this end, the strategy to be used will be to prioritise the dimensions that influence the others.

The dimensions *Technologies 4.0*, *Business Models* and *Systems Integration* are the ones that have a lower level of maturity in Company G. Thus, it was observed in Figure x that five dimensions (Human Resources, Knowledge Dissemination, Organisational Culture, Business Models and I4.0 Process) influence *Technologies 4.0,* and seven dimensions influence *Systems Integration* (Human Resources, Knowledge Dissemination, Organisational Culture, Business Models, TT Process, Technologies 4.0 and I4.0 Process). The *Business Models* dimension is an influencing dimension.

Thus, it is suggested that it is necessary to prioritize the *Business Models* dimension in the company action plans, for the reason that it is an influential dimension of BB4.0 and has a low level of maturity in the company. Subsequently, one should focus on the dimensions: Human Resources, Knowledge Dissemination, Organizational

Culture, TT Process, Technologies 4.0 and I4.0 Process. Consequently, other dimensions will benefit from the improvements obtained in these mentioned dimensions.

Subsequently, the feasibility and investment required for each of the actions presented must be analysed, in order to list which actions should be carried out.

The next step for Company G is to advance to Level V. To do so, it must continue investing in TT more and more, focusing on the search for and implementation of BB4.0.

8.4 CHAPTER CONSIDERATIONS

The purpose of this chapter was to present the results obtained in the application of the proposed model. It was possible to analyse which dimensions are influencing and influenced, the maturity level of Industry 4.0 of each interviewed company, and the individual diagnosis of the companies.

Of the ten dimensions analysed, four were deemed influential (Organizational Culture, Human Resources, Knowledge Dissemination and Business Models) and six were deemed influenced by these four mentioned. Of these four influencing dimensions, three of them refer to Technology Transfer and only one refers to Building Blocks 4.0. This proves the fact that TT is a strategic factor to leverage the concepts and technologies 4.0 in companies.

The objective of knowing which are the influencing and influenced dimensions is to be able to design strategies for companies to raise their I4.0 maturity level. Furthermore, this model is based on the influences between the dimensions.

The level of Industry 4.0 maturity of the companies studied was deemed generally satisfactory. Of the seven companies, two are at level I, one is at level II, and four are at level V.

The positioning of companies in the model allowed to prove the theoretical framework developed in this study, which states that there is a mutual influence between TT and R4.0. Furthermore, that Technology Transfer acts as a strategic factor for the evolutionary implementation of 4.0 concepts and technologies in organizations.

To conclude this chapter, suggestions of actions were made for each of the companies studied, based on the performance of each of the dimensions presented.

It was possible to verify that companies with a higher maturity level of Industry 4.0 had an acceptable maturity level of the influencing dimensions. Whereas companies with a lower maturity level of I4.0, presented a lack of maturity in the influencing dimensions.

Therefore, the present work will allow the participating industries to better understand the scenario they are trying to achieve, and thus be able to advance in this context.

9 FINAL CONSIDERATIONS

This study's overall objective was to develop an Industry 4.0 maturity model whereby Technology Transfer influences its maturity level so as to facilitate its implementation. So that this objective might be achieved, it was necessary to attain and respond to the specific objectives outlined in this study's introduction.

To answer the first specific objective, which would be the development of a Conceptual *Framework* that presents the influence between TT and I4.0, it was necessary to carry out a broad literature review. In this review, were presented: the concepts, components, impacts and barriers related to Industry 4.0; the mechanisms, concepts, impacts and barriers related to Technology Transfer; the maturity models found in the literature, both of Industry 4.0, as Technology Transfer; and finally, the union between TT and I4.0. From this review, it was elaborated the conceptual *framework*, which addressed that the TT becomes an ally to the Concepts and Technologies 4.0 for the implementation of Industry 4.0 in organizations.

In order to prove this conceptual *framework*, it was prepared the I4.0-TT Maturity Model. For this, it was first necessary to determine the components and dimensions of TT and I4.0, this being the second specific objective of the research. Both the components and the dimensions of TT and I4.0 were determined from the literature review. These components were classified into ten sub-dimensions (Human Resources, Human-Machine Interface, TT Process, I4.0 Process, Knowledge Dissemination, 4.0 Technologies, Organisational Culture, Business Models, TT Modes and System Integration) that fit the five dimensions presented in the Conceptual *Framework* (People, Process, Technology, Strategy and Organisation, and Integration).

The third specific objective is related to the model application. For this, it was developed a multicriteria model based on the DEMATEL method, where the dimensions have influence over each other, and thus it is possible to determine the

influence between TT and I4.0 and discover the maturity level of the company. This model was applied in seven companies located in Paraná. All of them intend to implement Industry 4.0 concepts and technologies. These companies were chosen for the reason that they have partnerships with the Federal Technological University of Paraná.

After the model application, it was possible to achieve the fourth specific objective: to analyze which are the I4.0 and TT dimensions that most influence the others and suffer influence. Of the ten dimensions analysed, four were considered influencing

(Organisational Culture, Human Resources, Knowledge Dissemination and Business Models), and six were considered influenced by these four mentioned. Of these four influencing dimensions, three of them refer to Technology Transfer and only one refers to *Building Blocks* 4.0. This proves the fact that TT is a strategic factor to leverage 4.0 concepts and technologies in companies. Knowing which are the influencing and influenced dimensions, it was possible to design strategies for companies to raise their I4.0 maturity level.

Subsequently, it was possible to identify in the proposed model at which Industry 4.0 maturity level the interviewed companies are located. Of the seven companies, two are at level I, one is at level II, and four are at level V.

The positioning of companies in the model enabled the conceptual *Framework* developed in this study to be confirmed, whereby it is explained that there is a mutual influence between TT and R4.0. Furthermore, that Technology Transfer acts as a strategic factor for the evolutionary implementation of 4.0 concepts and technologies within organizations.

Finally, it was developed a TT-I4.0 maturity diagnosis for the interviewed companies, being this the last specific research objective. Action suggestions were made for each of the companies studied, based on the performance of each of the dimensions presented. It was also outlined a strategy to prioritize the actions, based on the influencing dimensions.

In this sense, the answers to the specific objectives outlined for this study lead to the achievement of the overall objective. The construction of the I4.0-TT Maturity Model enabled the confirmation that Technology Transfer influences Industry 4.0 and vice-versa. TT can be seen as a strategy to facilitate the implementation of 4.0 concepts and technologies. Furthermore, it is a tool that can be used in organizations to analyse the current stage of the company and what is required to evolve to the next level.

Industry 4.0 is a very complex theme and companies still face difficulties in its implementation process. This study confirmed that Technology Transfer comes as a support to assist and facilitate the entire implementation process of this new industrial revolution. The evolution process of I4.0 and TT must be continuous and simultaneous, because both have a mutual influence relationship. Thus, this work filled an existing gap in the literature, and brings possibilities for new future studies.

9.1 SUGGESTIONS FOR FUTURE WORK

This research was limited by the absence of a visit to the companies for a more precise and pertinent diagnosis of the organization's actual situation. Only by assessing questionnaires, the analysis became more superficial. For future research, it is suggested to conduct an *in loco* visit to the companies studied.

Still, opportunities for the development of future works related to the theme were identified. They are the following:

- Apply the model built in other companies, making comparisons according to the size of the company and its field of activity;
- Apply the model built in different countries in order to make a comparison;
- Incorporate the built model into an Industry 4.0 *roadmap*;
- Apply the model in order to identify the existing TT and I4.0 barriers that hinder the implementation of this new industrial revolution.

REFERENCES

ABELE, E.; METTERNICH, J.; TISCH, M.; CHRYSSOLOURIS, G.; SIHN, W.; ELMARAGHY, W.; RANZ, F. Learning factories for research, education, and training. **Procedia CiRp**, v.32, p.1-6, 2015.

ADENLE, A. A.; AZADI, H.; ARBIOL, J. Global assessment of technological innovation for climate change adaptation and mitigation in developing world. **Journal of Environmental Management**, v. 161, p. 261-275, 2015.

AGBOOLA, J. I. Technological Innovation and Developmental Strategies for Sustainable Management of Aquatic Resources in Developing Countries. **Environmental Management**, v. 54, n. 6, p. 1237-1248, 2014.

AGGARWAL, P.; AGGARWAL, R. Examining perspectives and dimensions of clean development mechanism: A critical assessment vis-à-vis developing and least developed countries. **International Journal of Law and Management**, v. 59, n. 1, p. 82-101, 2017.

AGRAWAL, Ajay K. University-to-industry knowledge transfer: Literature review and unanswered questions. **International Journal of management reviews**, v. 3, n. 4, p. 285-302, 2001.

AHN, J.H.; CHANG, S.G. Assessing the contribution of knowledge to business performance: the KP3 methodology. **Decision Support Systems**, v. 36, n. 4, p. 403-416, 2004.

AIREHROUR, D.; GUTIERREZ, J.; RAY, S. K. Secure Routing for Internet of Things: A Survey. **Journal of Network Computers and Applications,** v.66, p.198-213, 2016.

AKDIL, K. Y.; USTUNDAG, A.; CEVIKCAN, E. Maturity and Readiness Model for Industry 4.0 Strategy. **In Industry 4.0: Managing the Digital Transformation,** p. 61-94, 2018.

AL-SAEED, Y.; EDWARDS, D. J.; SCAYSBROOK, S. Automating construction manufacturing procedures using BIM digital objects (BDOs): Case study of knowledge transfer partnership project in UK. **Construction Innovation.** 2020.

AMESSE, F.; COHENDET, P. Technology transfer revisited from the perspective of the knowledge- based economy. **Research Policy**, v.30, p.459-1478, 2001.

ANSARI, F.; HOLD, P.; KHOBREH, M. A knowledge-based approach for representing jobholder profile toward optimal human-machine collaboration in cyber physical production systems. **CIRP Journal of Manufacturing Science and Technology**, v. *28*, p. 87-106, 2020.

ANTKOWIAK, D.; LUETTICKE, D.; LANGER, T.; THIELE, T.; MEISEN, T.; JESCHKE, S. Cyber-Physical Production Systems: A Teaching Concept in Engineering Education. **Proceedings - 2017 6th IIAI International Congress on Advanced Applied Informatics**, **IIAI-AAI,** p. 681-686, 2017.

ARGOTE, L. Organizational learning: Creating, retaining and transferring knowledge. **Boston: Kluwer,** (pp. 1-212), 1999.

ARGOTE, L.; INGRAM, P. Knowledge transfer: a basis for competitive advantage in firms. **Organizational Behavior and Human Decision Processes**, v. 82, n. 1, p. 150-169, 2000.

ARIF, M.; AL ZUBI, M.; GUPTA, A. D.; EGBU, C.; WALTON, R. O.; ISLAM, R. Knowledge sharing maturity model for Jordanian construction sector. **Engineering, Construction and Architectural Management**, 2017.

ARNOLD, C.; KIEL, D.; VOIGT, K.I. How the Industrial Internet of Things changes business models in different manufacturing industries. **International Journal of Innovation Management,** v.20, n.08, p.1640015-1-1640015-25, 2016.

ARTIFICE, A. et al. Computational Model for Knowledge Transfer Skills in Industry 4.0 in an Enhanced and Effective Way. In: **ASME International Mechanical Engineering Congress and Exposition**. American Society of Mechanical Engineers, 2019.

ARVANITIS, S.; KUBLI, U.; WOERTER, M. University-industry knowledge and technology transfer in Switzerland: what university scientists think about co-operation with private enterprises. **Research Policy**, v.37, n. 10, p. 1865-1883, 2008.

ARVANITIS, S.; SYDOW, N.; WOERTER, M. Is there any impact of university-industry knowledge transfer on innovation and productivity? An empirical analyses based on Swiss firm data. **Review of Industrial Organization**, v. 32, n. 2, p. 77-94, 2008.

ARZA, V. Channels, benefits and risks of public-private interactions for knowledge transfer: conceptual framework inspired by Latin America. **Science and Public Policy**, v. 37, n. 7, p. 473-484, 2010.

ATZORI, L.; ANTONIO, I.; GIACOMO, M. The Internet of Things: A survey. **Computer Networks, v.** 54, n.15, p. 2787-2805, 2010.

AUTIO, E.; LAAMANEN, T. Measurement and evaluation of technology transfer: Review of technology transfer mechanisms and indicators. **Technology Management**, v.10, n.7/8, p.643-664, 1995.

AZAM, M. M. Climate Change Resilience and Technology Transfer: The Role of Intellectual Property. **Nordic Journal of International Law**, v. 80, n. 4, p. 485-505, 2011.

BABICEANU, R. F.; SEKER, R. Big Data and Virtualization for Manufacturing Cyber-Physical Systems: A Survey of the Current Status and Future Outlook. **Computers in Industry,** v.81, p.128-137, 2016.

BABICEANU, R. F.; SEKER, R. Big data and virtualization for manufacturing cyber-physical systems: a survey of the current status and future outlook. **Computers in Industry**, v.81, p.128-137, 2016.

BAHRIN, M.A.K.; OTHMAN, M.F.; AZLI, N.H.N.; TALIB, M.F. Industry 4.0: a review on industrial automation and robotic. **Jurnal Teknologi,** v.78, p.6-13, 2016.

BARAKI, Y. A.; BRENT, A. C. Technology transfer of hand pumps in rural communities of Swaziland: Towards sustainable project life cycle management. **Technology in Society**, v. 35, n. 4, p. 258-266, 2013.

BARBOSA, G. F.; SHIKI, S. B.; DA SILVA, I. B. R&D roadmap for process robotization driven to the digital transformation of the industry 4.0. **Concurrent Engineering Research and Applications**, v. *28*, n.4, p. 290-304, 2020.

BARLETTA, F.; YOGUEL, G.; PEREIRA, M.; RODRIGUEZ, S. Exploring scientific productivity and transfer activities: evidence from Argentinean ICT research groups. **Research Policy**, v.46, n. 8, p. 1361-1369, 2017.

BASU, A. Grasping climate technology transfer: A brief discussion on Indian practice. **Journal of Intellectual Property Rights**. v. 23, p. 51-59, 2018.

BATTISTELLA, Cinzia; DE TONI, Alberto F.; PILLON, Roberto. Inter-organisational technology/knowledge transfer: a framework from critical literature review. **The Journal of Technology Transfer**, v. 41, n. 5, p. 1195-1234, 2016.

BAUR, C.; WEE, D. Manufacturing's Next Act? **McKinsey & Company**, 2015.

BEDOLLA, J. S.; D'ANTONIO, G.; CHIABERT, P. A Novel Approach for Teaching IT Tools within Learning Factories. **Procedia Manufacturing**, v.9, p.175-181, 2017.

BEER, C.; SECUNDO, G.; PASSIANTE, G.; SCHUTTE, C. S. A mechanism for sharing best practices between university technology transfer offices. **Knowledge Management Research & Practice**, v.15, n.4, p.523-532, 2017.

BEKKERS, R.; BODAS, I. Analyzing knowledge transfer channels between universities and industry: to what degree do sectors also matter? **Research Policy**, v. 37, n. 10, p. 1837-1853, 2008.

BELMONTE, S.; ESCALANTE, K. N.; FRANCO, J. Shaping changes through participatory processes: Local development and renewable energy in rural habitats. **Renewable and Sustainable Energy Reviews**, v. 45, p. 278-289, 2015.

BEN-ARI, Mordechai; MONDADA, Francesco. **Elements of robotics**. Cham, Switzerland: Springer International Publishing, 2018.

BENEDICT, N.; SMITHBURGER, P.; DONIHI, A. C.; EMPEY, P.; KOBULINSKY, L.; SEYBERT, A.; WATERS, T.; DRAB, S.; LUTZ, J.; FARKAS, D.; MEYER, S. Blended Simulation Progress Testing for Assessment of Practice Readiness. **American journal of pharmaceutical education**, v.81, n.1, p.14, 2017.

BENEŠOVÁ, A.; TUPA, J. Requirements for education and qualification of people in industry 4.0. **Procedia Manufacturing**, v.11, p. 2195-2202, 2017.

BERDAL, Q.; PACAUX-LEMOINE, M. P.; TRENTESAUX, D.; CHAUVIN, C. Human-machine cooperation in self-organized production systems: A point of view. **International Workshop on Service Orientation in Holonic and Multi-Agent Manufacturing,** p. 123-132, 2018.

BLIZNETS, I.; KARTSKHIYA, A.; SMIRNOV, M. Technology transfer in digital era: legal environment. **Journal of History Culture and Art Research**, v. 7, n. 1, p. 354-363, 2018.

BLOHMKE, J. Technology complexity, technology transfer mechanisms and sustainable development. **Energy for Sustainable Development**, v. 23, p. 237-246, 2014.

BLÜMEL, E. Global Challenges and Innovative Technologies Geared Toward New Markets: Prospects for Virtual and Augmented Reality. **VARE**, p. 4-13, 2013.

BOGLE, I. D. L. A Perspective on Smart Process Manufacturing Research Challenges for Process Systems Engineers. **Engineering,** v.3, n.2, p.161-165, 2017.

BOZEMAN, B. Technology transfer and public policy: a review of research and theory. **Research Policy**, v. 29, n. 4/5, p. 627-655, 2000.

BOZEMAN, B.; RIMES, H.; YOUTIE, J. The evolving state-of-the-art in technology transfer research: Revisiting the contingent effectiveness model. **Research Policy**, v.44, n.1, p.34-49, 2015.

BRANGER, J.; PANG, Z. Automated Home to Sustainable, Healthy and Manufacturing Home: A New Story Enabled by the Internet-of-Things and Industry 4.0. **Journal of Management Analytics,** v.2, n.4, p.314-332, 2015.

BRÖRING, A.; SCHMID, S.; SCHINDHELM, C. K.; KHELIL, A.; KÄBISCH, S.; KRAMER, D.; TENIENTE, E. Enabling IoT ecosystems through platform interoperability. **IEEE Softwer**, v.34, n.1, p.54-61, 2017.

BRUCKMAN, V.; HARUTHAITHANASAN, M.; MILLER, R.; TERADA, T.; BRENNER, A.-K.; KRAXNER, F.; FLASPOHLER, D. Sustainable forest bioenergy development strategies in Indochina: Collaborative effort to establish regional policies. **Forests**, v.

9, n. 4, p. 223, 2018.

BYRNE, G.; AHEARNE E.; COTTERELL, M.; MULLANY, B.; O'DONNELL, G. E.; SAMMLER, F. High Performance Cutting (HPC) in the New Era of Digital Manufacturing: A Roadmap. **Procedia CIRP,** v.46, p.1-6, 2016.

CAMP, S. M.; SEXTON, D. L. Technology Transfer and Value creation: Extending the theory beyond information exchange. **The Journal of Technology Transfer**, v. 17, n. 2-3, p. 68-76, 1992.

CARAYANNIS, E.G; CAMPBELL, D.F.J. Mode 3" and "quadruple helix": toward a 21st century fractal innovation ecosystem. **International Journal of Technology Management**, v.46, p.201-234, 2009.

CARVALHO, I. V.; CUNHA, N. C. V. Proposal of a technology transfer model for Brazilian public universities. **Congresso Latino-Iberoamericana de Gestão de Tecnologia,** v.15, p.1-18, 2013.

CARVALHO, N.; CHAIM, O.; CAZARINI, E.; GEROLAMO, M. Manufacturing in the fourth industrial revolution: a positive prospect in sustainable manufacturing. **Procedia Manufacturing**, v.21, p.671-678, 2018.

CEDEÑO, J. M. V.; PAPINNIEMI, J.; HANNOLA, L.; DONOGHUE, I. Developing smart services by internet of things in manufacturing business. **LogForum,** v.14, n.1, p.59-71, 2018.

CHANG, B.; CHANG, C.; WU, C. Fuzzy DEMATEL method for developing supplier selection criteria. **Expert Systems with Applications**, v.38, n.3, p.1850-1858, 2011.

CHEGE, M. S.; WANG, D.; SUNTU, S. L.; BISHOGE, O. K. Influence of technology transfer on performance and sustainability of standard gauge railway in developing countries. **Technology in Society**, v. 26, p. 79-92, 2019.

CHEN, T.; LIN, Y.C. Feasibility evaluation and optimization of a smart manufacturing system based on 3D printing: a review. **International Journal of Intelligent Systems,** v.32, n.4, p. 394-413, 2017.

CHENG, G.; LIU, L.; QIANG, X.; LIU, Y. Industry 4.0 Development and Application of Intelligent Manufacturing. **Proceedings of 2016 International Conference on Information Systems and Artificial Intelligence**, p.407-410, 2016.

CHOI, S.; KIM, B. H.; DO NOH, S. A diagnosis and evaluation method for strategic planning and systematic design of a virtual fac- tory in smart manufacturing systems. **International Journal of Precision Engineering and Manufacturing**, v.16, n.6, p. 1107-1115, 2015.

CHONG, Li; RAMAKRISHNA, Seeram; SINGH, Sunpreet. A review of digital manufacturing-based hybrid additive manufacturing processes. **The International**

Journal of Advanced Manufacturing Technology, v. 95, n. 5-8, p. 2281-2300, 2018.

CHRISTIAN, E.; GLAESNER, K.; ALAIN, P.; JOSIP, S. Advances in 3D Measurement Data Management for Industry 4.0. **Procedia Manufacturing**, v.11, p.1335-1342, 2017.

CHRISTIANS, A.; LIEPIN, M. The consequences of digitalization for German civil law from the national legislator's point of view. **Zeitschrift Geistiges Eigentum/ Intellectual Property Journal,** v.9, n.3, p.331-339, 2017.

CIANFANELLI, E.; GORETTI, G.; TUFARELLI, M. Reverse engineering and digital archives as a Resource for Practical Craft-Based Manufacturing Process. **International Conference on Applied Human Factors and Ergonomics**, p. 280-289, 2018.

CORMICAN, K.; O'CONNOR, M. Technology transfer for product life cycle extension: A model for successful implementation. **International Journal of Innovation and Technology Management**, v. 6, n. 03, p. 265-282, 2009.

CORRADINI, G.; BROTTO, L.; CICCARESE, L.; PETTENELLA, D. An overview of Italian participation in afforestation and reforestation projects under the Clean Development Mechanism**. iForest - Biogeosciences and Forestry**, v. 9, n. 5, p. 720-728, 2016.

CORREA, Carlos M. Innovation and Technology Transfer of Environmentally Sound Technologies: The Need to Engage in a Substantive Debate. **Review of European, Comparative & International Environmental Law,** v. 22, n. 1, p. 54-61, 2013.

DA SILVA, Vander Luiz; KOVALESKI, João Luiz; PAGANI, Regina Negri. Technology transfer in the supply chain oriented to industry 4.0: a literature

DA XU, L.; HE, W.; LI, S. Internet of things in industries: a survey. **IEEE Trans. Ind. Inf**, v.10, n.4, p.2233-2243, 2014.

DAIS, S. Industrie 4.0 - Anstoß, Vision, Vorgehen. In: Bauernhansl T, ten Hompel M, Vogel Heuser B (eds) Industrie 4.0 in Produktion, Automatisierung und Logistik: Anwendung. Technologien. Migration. **Springer, Wiesbaden**, p.625-634, 2014.

DASGUPTA, P. TANEJA, N. Low Carbon Growth: An Indian Perspective on Sustainability and Technology Transfer. **Problemy Ekorozwoju - Problems of Sustainable Development**, v. 6, p. 65-74, 2011.

DAVIES, A.; FIDLER, D.; GORBIS, M. Future Work Skills 2020, **Available at Institute for the Future for University of Phoenix Research Institute**, p. 540, 2011

DAVIS, J.; EDGAR, T.; GRAYBILL, R.; KORAMBATH, P.; SCHOTT, B.; SWINK, D.;

WANG, J.; WETZEL, J. Smart Manufacturing, **Annual review of chemical and biomolecular engineering.** v.6, p. 141-160, 2015.

DAVIS, J.; EDGAR, T.; PORTER, J.; BERNADEN, J.; SARLI, M. Smart manufacturing, manufacturing intelligence and demand-dynamic performance. **Comput Chem Eng,** v.47, p.145-156, 2012.

DE CONINCK, Heleen; HAAKE, Frauke; VAN DER LINDEN, Nico. Technology transfer in the clean development mechanism. **Climate policy**, v. 7, n. 5, p. 444-456, 2007.

DEBACKERE, K.; ANDERSEN, B.; DVORAK, I.; ENKEL, E.; KRUGER, P.; MALMQVIST, H.; PLECKAITIS, A.; REHN, A.; SECALL, S.; VERMEULEN, E.; WELLEN, D. Boosting Open Innovation and Knowledge Transfer in the EU: Independent Expert Group Report on Open Innovation and Knowledge Transfer, **European Commission**, Brussels, 2014.

DEBACKERE, K.; VEUGELERS, R. The role of academic technology transfer organizations in improving industry science link. **Research Policy**, v. 34, n. 3, p. 321-342, 2005.

DJUMALIEVA, J.; SLEEMAN, C. Which digital skills do you really need?. **Nesta**, 2018.

DOMBROWSKI, U.; WAGNER, T. **Procedia CIRP,** v.17, p.100-105, 2014.

DOUKAS, H.; KARAKOSTA, C.; PSARRAS, J. RES technology transfer within the new climate regime: A "helicopter" view under the CDM. **Renewable and Sustainable Energy Reviews**, v. 13, n. 5, p. 1138-1143, 2009.

DUTRÉNIT, G.; ARZA, V. Channels and benefits of interactions between public research organisations and industry: comparing four Latin American countries. **Science and Public Policy**, v. 37 n. 7, p. 541-553, 2010.

EATON, D. Technology and Innovation for a Green Economy. **Review of European, Comparative & International Environmental Law**, v. 22, n. 1, p. 62-67, 2013.

EFSTATHIADES, Andreas; TASSOU, Savvas; ANTONIOU, Antonios. Strategic planning, transfer and implementation of Advanced Manufacturing Technologies (AMT). Development of an integrated process plan. **Technovation**, v. 22, n. 4, p. 201-212, 2002.

EITZEL, M. V.; HOVE, E. M.; SOLERA, J.; MADZORO, S.; CHANGARARA, A.; NDLOVU, D.; GWATIPEDZA, S.; MHIZHA, M.; NDLOVU, M. Sustainable development as successful technology transfer: Empowerment through teaching, learning, and using digital participatory mapping techniques in Mazvihwa, Zimbabwe. **Development Engineering**, v. 3, p. 196-208, 2018.

EROL, S.; JÄGER, A.; HOLD, P.; OTT, K.; SIHN, W. **Procedia CIRP,** v.54, p.13-18, 2016.

ESCALANTE, K. N.; BELMONTE, S.; GEA, M. D. Determining factors in process of socio-technical adequacy of renewable energy in Andean Communities of Salta, Argentina. **Renewable and Sustainable Energy Reviews**, v. 22, p. 275-288, 2013.

ESMAEILIAN, Behzad; BEHDAD, Sara; WANG, Ben. The evolution and future of manufacturing: A review. **Journal of Manufacturing Systems**, v. 39, p. 79-100, 2016.

ETZKOWITZ, H.; LEYDESDORFF, L. The dynamics of innovation: from national systems and 'mode 2' to a tri- ple helix of university-industry-government relations. **Research Policy**, v.29, p.109-123, 2000.

FANG C.; LIU, X.; PARDALOS, P. M.; PEI, J. Optimization for a three-stage production system in the internet of things: procurement, production and product recovery, and acquisition, **The International Journal of Advanced Manufacturing Technology**, v.83, n.5-8, p.689-710, 2016.

FASEHUN, A. O. The War on Climate Change: Ushering in Sustainable Development to VDCs through a Technical Capacity-building Facility. **Environmental Claims Journal**, v. 27, n. 3, p. 196-225, 2015.

FENG, S. C.; BERNSTEIN, W. Z.; HEDBERG JR, T. Towards Knowledge Management for Smart Manufacturing. **Journal of Computing and Information Science in Engineering,** v.17, n.3, p.1-40, 2017.

FERREIRA, J. J.; FERNANDES, C.; RATTEN, V. Environmental-related patent technology transfer effectiveness: A comparison between Portugal and Australia using OECD data. **World Journal of Entrepreneurship, Management and Sustainable Development**, v. 14, n. 3, p. 206-221, 2018.

FLAMOS, A. The clean development mechanism-catalyst for wide spread deployment of renewable energy technologies? or misnomer? **Environment, Development and Sustainability**, v. 12, n. 1, p. 89-102, 2010.

FLAMOS, A.; BEGG, K. Technology transfer insights for new climate regime. **Environment, Development and Sustainability**, v. 12, n. 1, p. 19-33, 2010.

FLAMOS, A.; GEORGALLIS, P. G.; PSARRAS, J. Bioenergy Options in the Industrialized and Developing World and Opportunities for the Clean Development Mechanism. **International Journal of Green Energy**, v. 7, n. 6, p. 647-661, 2010.

FOIDL, H.; FELDERER, M. Research Challenges of Industry 4.0 for Quality Management. **Innovations in Enterprise Information Systems Management and Engineering, Springer**, p. 121-137, 2016.

FREEMAN, C. The 'National System of Innovation' in historical perspective. **Cambridge Journal of Economics**, v.19, p. 5-24, 1995.

FREY, C.B.; OSBORNE, M.A. The future of employment: how susceptible are jobs to computerization? **Technology Forecasting Social Change,** v.114, p.254-280, 2017.

FU, Y.; DING, J.; WANG, H.; WANG, J. Two-objective stochastic flow-shop scheduling with deteriorating and learning effect in Industry 4.0-based manufacturing system. **Applied Soft Computing**, v.68, p.847-855, 2018.

FU, S.; LI, Z.; LIU, K.; DIN, S.; IMRAN, M.; YANG, X. Model Compression for IoT Applications in Industry 4.0 via Multiscale Knowledge Transfer. **IEEE Transactions on Industrial Informatics,** v. *16*, n. 9, p. 6013-6022, 2019.

FUENTES, C.; DUTRÉNIT, G. Best channels of academia-industry interaction for long-term benefit. **Research Policy**, v. 41, n. 9, p. 1666-1682, 2012.

GABUS, A.; FONTELA, E. World problems, an invitation to further thought within the framework of DEMATEL. Switzerland, Geneva, 1972.

GALLAGHER, K. S. Limits to leapfrogging in energy technologies? Evidence from the Chinese automobile industry. **Energy Policy**, v. 34, n. 4, p. 383-394, 2006.

GANZARAIN, Jaione; ERRASTI, Nekane. Three stage maturity model in SME's toward industry 4.0. **Journal of Industrial Engineering and Management (JIEM)**, v. 9, n. 5, p. 1119-1128, 2016.

GARBELLANO, S.; DA VEIGA, M. R. Dynamic capabilities in Italian leading SMEs adopting industry 4.0. **Measuring Business Excellence,** v. 23, n.4, p.472-483, 2019.

GARCIA, M.L.; BRAY, O.H. Fundamentals of Technology Roadmapping (No. SAND-97-0665). **Sandia National Labs, Albuquerque, NM (United States)**, 1997.

GEORGE, M.; OSMAN, A. S. S.; GEORGE, A. R.; HUSSIN, H. Protecting the Malacca and Singapore Straits from Ships' Atmospheric Emissions through the Implementation of MARPOL Annex VI. **The International Journal of Marine and Coastal Law**, v. 32, n. 1, p. 95-137, 2017.

GERSTLBERGER, W. Regional innovation systems and sustainability-selected examples of international discussion. **Technovation**, v. 24, n. 9, p. 749-758, 2004.

GILCHRIST, A. Industry 4.0: The Industrial Internet of Things. Springer, Heidelberg, 2016.

GILSING, V. Differences in technology transfer between science-based and development-based industries: Transfer mechanisms and barriers. **Technovation,** v. 31, n. 12, p. 638-647, 2011.

GISSELQUIST, D.; GRETHER, J.M. An argument for deregulating the transfer of agriculture technologies to developing countries. **World Bank Econ**, v.14, n.10, p.111-128, 2000.

GJELDUM, N.; MLADINEO, M.; VEŽA, I. Transfer of model of innovative smart factory to Croatian economy using lean learning factory. **Procedia CIRP**, v.54, p.158-163, 2016.

GNIMPIEBA, Z. D. R.; NAIT-SIDI-MOH, A.; DURAND, D.; FORTIN, J. Using internet of things technologies for a collaborative supply chain: application to tracking of pallets and containers. **Procedia Computer Science**, v.56, p.550-557, 2015.

GÖKALP, Ebru; ŞENER, Umut; EREN, P. Erhan. Development of an assessment model for industry 4.0: industry 4.0-MM. In: **International Conference on Software Process Improvement and Capability Determination**. Springer, Cham, p. 128-142, 2017.

GORECKY, D.; SCHMITT, M.; LOSKYLL, M.; ZÜHLKE, D. Human-machine-interaction in the industry 4.0 era. **IEEE International Conference on Industrial Informatics IEEE,** p.289-294, 2014.

GRIMALDI, R.; KENNEY, M.; SIEGEL, D.S.; WRIGHT, M. 30 Years after Bayh-dole: reassessing academic entrepreneurship. **Research Policy**, v.40, p.1045-1057, 2011.

GUPTA, P.; SEETHARAMAN, A.; RAJ, J. R. The Usage and Adoption of Cloud Computing by Small and Medium Businesses. **International Journal of Information Management,** v.33, p.861-874, 2013.

HAJRIZI, E.; STAPLETON, L.; KOPACEK, P. The Contribution of UBT for the sustainable development of Kosova. **IFAC Proceedings Volumes**, v. 43, n. 25, p. 47-52, 2010.

HAMID, N.; SALIM, J. A conceptual framework of knowledge transfer in Malaysia e-government IT outsourcing: an integration with transactive memory system (TMS). **IJCSI International Journal of Computer Science Issues**, v. 8, n.5, p. 51-64, 2011.

HANNOLA, L.; RICHTER, A.; RICHTER, S.; STOCKER, A. Empowering production workers with digitally facilitated knowledge processes-a conceptual framework. **International Journal of Production Research**, v.56, n.14, p.4729-4743, 2018.

HECKLAU, F.; GALEITZKE, M.; FLACHS, S.; KOHL, H. Holistic approach for human resource management in industry 4.0. **Procedia CIRP,** v.54, 2016.

HERMANN, M.; PENTEK, T.; OTTO, B. Design Principles for Industrie 4.0 Scenarios. **Proceedings of the Annual Hawaii International Conference on System Sciences**, p. 3928-3937, 2016.

HOGNELID, P.; KALLING, T. Internet of things and business models. **In**

standardization and innovation in information technology (SIIT), IEEE 9th International Conference on, IEEE, p. 1-7, 2015.

HUANG, X.; HSIEH, J.J.; HE, W. Expertise dissimilarity and creativity: the contingent roles of tacit and explicit knowledge sharing**. Journal of Applied Psychology**, v. 99, n. 5, p. 816-830, 2014.

HUSSAIN, M. I. Internet of things: challenges and research opportunities. **CSI Trans, ICT,** v.5, n.1, p.87-95, 2017.

IGNATAVIČIUS, R.; TVARONAVIČIENĖ, M.; PICCINETTI, L. Sustainable development through technology transfer networks: Case of Lithuania. **Journal of Security and Sustainability Issues**, v. 4, p. 261-267, 2015.

ILIE, Iulian; GHEORGHE, Gheorghe I. Role of technological transfer for mems & nems in intelligent manufacturing process. **Romanian Review Precision Mechanics, Optics & Mechatronics**, n. 45, p. 175, 2014.

INBAL, A. B.; TZACHOR, A. National policy and SMEs in technology transfer: the case of Israel. **Climate Policy**, v. 15, n. 1, p. 88-102, 2013.

ISMAIL, M. Conceptualizing knowledge transfer between expatriates and host country nationals: the mediating effect of social Capital. **Cogent Business and Management**, v. 2, n. 1, p. 1-16, 2015.

ISO, 2015. ISO/IEC 33004 - Information technology - Process assessment - Requirements for process reference, process assessment and maturity models. 2015.

ITUARTE, I. F.; CHEKUROV, S.; TUOMI, J.; MASCOLO, J. E.; ZANELLA, A.; SPRINGER, P.; PARTANEN, J. Digital manufacturing applicability of a laser sintered component for automotive industry: A case study. **Rapid Prototyping Journal**, 2018.

IVANOVA, I. Quadruple helix systems and symmetry: a step towards helix innovation system Classification. **Journal of the Knowledge Economy**, v.5, p.357-369, 2014.

IVEZIC, N.; KULVATUNYOU, B.; SRINIVASAN, V. On Architecting and Composing Through-life Engineering Information Services to Enable Smart Manufacturing. **Procedia CIRP**, v.22, p. 45-52, 2014.

JARA, Antonio J.; PARRA, María Concepción; SKARMETA, Antonio F. Marketing 4.0: A new value added to the Marketing through the Internet of Things. In: **2012 Sixth International Conference on Innovative Mobile and Internet Services in Ubiquitous Computing**. IEEE, 2012. p. 852-857.

JASON, S.; CURIEL, A. S.; LIDDLE, D.; CHIZEA, F.; LELOGLU, U. M.; HELVACI, M.; BEKHTI, M.; BENACHIR, D.; BOLAND, L.; GOMES, L.; SWEETING, M. Capacity building in emerging space nations: Experiences, challenges and benefits. **Advances**

in Space Research, v. 46, n. 5, p. 571-581, 2010.

JIANG, R.; KLEER, R.; PILLER, F. T. Predicting the future of additive manufacturing: A Delphi study on economic and societal implications of 3D printing for 2030. **Technology Forecasting Social Change**, v.117, p.84-97, 2017.

KAGERMANN, H.; ANDERL, R.; GAUSEMEIER, J.; SCHUH, G.; WAHLSTER, W. Industrie 4.0 in a Global Context. **Munich: Herbert Utz Verlag**, 2016.

KAGERMANN, H.; WAHLSTER, W.; HELBIG, J. Recommendations for Implementing the Strategic Initiative INDUSTRIE 4.0, München, 2013.

KAGERMANN, H.; WAHLSTER, W.; HELBIG, J. Recommendations for Implementing the Strategic Initiative Industrie 4.0. **Industrie 4.0 Working Group, Germany**, 2013.

KAHN, Kenneth B.; CASTELLION, George; GRIFFIN, Abbie (Ed.). **The PDMA handbook of new product development**. Hoboken, NJ: Wiley, 2005.

KAMBLE, Sachin S.; GUNASEKARAN, Angappa; GAWANKAR, Shradha A. Sustainable Industry 4.0 framework: A systematic literature review identifying the current trends and future perspectives. **Process Safety and Environmental Protection**, v. 117, p. 408-425, 2018.

KAMIGAKI, T. ObjectOriented RFID with IoT: a design concept of information systems in manufacturing. **Electronics,** v.6, n.1, p.14, 2017.

KANG, H. S.; JU, Y. L.; SANG, S. C.; HYUN, K; JUN, H. P.; JI, Y. S.; BO, H. K.; SANG, D. N. Smart Manufacturing: Past Research, Present Findings, and Future Directions. **International Journal of Precision Engineering and Manufacturing-Green Technology,** v.3, n.1, p.111-128, 2016.

KANG, M. J.; PARK, J. Analysis of the partnership network in the clean development mechanism. **Energy Policy**, 52, pp. 543-553, 2013.

KARAKOSTA, C.; DOUKAS, H.; PSARRAS, J. Carbon market and technology transfer: statistical analysis for exploring implications. **International Journal of Sustainable Development & World Ecology**, v. 19, n. 4, p. 311-320, 2012.

KARAKOSTA, C.; DOUKAS, H.; PSARRAS, J. CDM sustainable technology transfer grounded in participatory in-country processes in Israel. **International Journal of Sustainable Society**, v. 3, n. 3, p. 225-242, 2011.

KARAKOSTA, C.; DOUKAS, H.; PSARRAS, J. EU-MENA energy technology transfer under the CDM: Israel as a frontrunner? **Energy Policy**, v. 38, n. 5, p. 2455- 2462, 2010a.

KARAKOSTA, C.; DOUKAS, H.; PSARRAS, J. Shaping sustainable development strategies in Chile through CDM. **International Journal of Climate Change Strategies and Management**, v. 1, n. 4, p. 382-399, 2009.

KARAKOSTA, C.; DOUKAS, H.; PSARRAS, J. Technology transfer through climate change: Setting a sustainable energy pattern. **Renewable and Sustainable Energy Reviews,** v. 14, n. 6, p. 1546-1557, 2010b.

KARAKOSTA, C.; PSARRAS, J. Facilitating sustainable development in Chile: a survey of suitable energy technologies. **International Journal of Sustainable Development & World Ecology**, v. 16, n. 5, p. 322-331, 2009.

KARRE, H.; HAMMER, M.; KLEINDIENST, M.; RAMSAUER, C. Transition towards an Industry 4.0 state of the LeanLab at Graz University of Technology. **Procedia manufacturing**, v.9, p.206-213, 2017.

KATHURIA, Vinish. Technology transfer for GHG reduction: A framework with application to India. **Technological Forecasting and Social Change**, v. 69, n. 4, p. 405-430, 2002.

KAYNAK, O. The Exhilarating Journey from Industrial Electronics to Industrial Informatics. **IEEE Transactions on Industrial Informatics,** v.1, n.2, p.73, 2005.

KENNEDY, Matthew; BASU, Biswajit. Overcoming barriers to low carbon technology transfer and deployment: An exploration of the impact of projects in developing and emerging economies. **Renewable and Sustainable Energy Reviews**, v. 26, p. 685-693, 2013.

KHAITAN, SIDDHARTHA, K.; JAMES, D. M. Design Techniques and Applications of Cyberphysical Systems: A Survey. **IEEE Systems Journal,** v.9, n.2, p.350-365, 2015.

KHAN, A.; TUROWSKI, K. A survey of current challenges in manufacturing industry and preparation for Industry 4.0, Proceedings of the First International Scientific Conference Intelligent Information Technologies for Industry, **Springer International Publishing**, p. 15-26, 2016.

KHAN, J.; HALEEM, A.; HUSAIN, Z. Barriers to technology transfer: a total interpretative structural model approach. **International Journal of Manufacturing Technology and Management**, v. 31, n. 6, p. 511-536, 2017.

KOCH, C.; BLIND, K. Towards Agile Standardization: Testbeds in Support of Standardization for the IIoT. **IEEE Transactions on Engineering Management**, v. 68, n. 1, p. 59-74, 2020.

KOCH, P. J.; VAN AMSTEL, M. K.; DEĘBSKA, P.; THORMANN, M. A.; TETZLAFF, A. J.; BØGH, S.; CHRYSOSTOMOU, D. A Skill-based Robot Co-worker for Industrial Maintenance Tasks, **Procedia Manuf**, v.11, p.83-90, 2017.

KOCH, V.; KUGE, S.; GEISSBAUER, R.; SCHRAUF, S. Industry 4.0: Opportunities and Challenges of the Industrial Internet. **Strategy & PwC**, 2014.

KOCIAN, J.; TUTSCH, M.; OZANA, S.; KOZIOREK, J. Application of modeling and simulation techniques for technology units in industrial control. In: **Frontiers in Computer Education**. Springer, Berlin, Heidelberg, 2012. p. 491-499.

KOEFOED, Michael; BUCKLEY, Chris. Clean technology transfer: a case study from the South African metal finishing industry, 2000-2005. **Journal of Cleaner Production**, v. 16, n. 1, p. S78-S84, 2008.

KOHLEGGER, M.; MAIER, R.; THALMANN, S. Understanding maturity models results of a structured content analysis. **Proceedings of the I-KNOW 2009-9th International Conference on Knowledge Management and Knowledge Technologies and Proceedings of I-SEMANTICS 2009**, Graz, Austria, p. 51-61, 2009.

KOSIERADZKA, A. Maturity model for production management. **Procedia Eng**, v.182, p.342-349, 2007.

KREILING, Laura; BOUNFOUR, Ahmed. A practice-based maturity model for holistic TTO performance management: development and initial use. **The Journal of Technology Transfer**, pp. 1-30, 2019.

KRUCKENBERG, L. J. North-South partnerships for sustainable energy: Knowledge-power relations in development assistance for renewable energy**. Energy for Sustainable Development**, v. 29, p. 91-99, 2015a.

KRUCKENBERG, L. J. Renewable energy partnerships in development cooperation: Towards a relational understanding of technical assistance. **Energy Policy**, v. 77, p. 11-20, 2015b.

KRÜCKHANS, B.; MORLOCK, F.; PRINZ, C.; FREITH, S.; KREIMEIER, D.; KUHLENKÖTTER, B. Learning Factories qualify SMEs to operate a smart factory. **COMA'16 Proceedings: International Conference on Competitive Manufacturing,** v.27-29, p. 457-460, 2016.

KRUGER, S.; STEYN, A. A. Enhancing technology transfer through entrepreneurial development: practices from innovation spaces. **Journal of Technology Transfer**, v. *45*, n. 6, p.1655-1689, 2020.

KUKKO, M. Knowledge sharing barriers in organic growth: A case study from a software company. **The Journal of High Technology Management Research**, 2013, v. 24, n. 1, p. 18-29.

KUMAR, Akshi; NAYYAR, Anand. si 3-Industry: A Sustainable, Intelligent, Innovative, Internet-of-Things Industry. In: **A Roadmap to Industry 4.0: Smart Production, Sharp Business and Sustainable Development**. Springer, Cham, p. 1-21, 2020.

KUSIAK, Andrew. Smart manufacturing. **International Journal of Production Research**, v. 56, n. 1-2, pp. 508-517, 2018.

KWAK Y.H.; IBBS C.W. Project management process maturity (PM) 2 model. **Journal of Management Engineering**, v.18, n.3, p.150-155, 2002.

LAMB, F.; ARLETT, C.; DALES, R.; DITCHFIELD B.; PARKIN, B. Engineering graduates for industry. **The Royal Academy of Engineering**, 2010.

LANZA, G.; HAEFNER, B.; KRAEMER, A. Optimization of selec- tive assembly and adaptive manufacturing by means of cyber- physical system based matching. **CIRP Annals - Manufacturing Technology**, v. 64, n. 1, p. 399-402, 2015.

LASI, H.; FETTKE, P.; KEMPER, G.; FELD, T.; HOFFMANN, M. Industry 4.0: Bedarfssog und Technologiedruck als Treiber der vierten Industrillen Revolution. **The International Journal of Wirtschaftsinformatik**, v.56, p.261-264, 2014.

LASRADO, L.A. **Set-Theoretic Approach to Maturity Models**, PhD Series, n. 15. ISBN 9788793579774. Copenhagen Business School (CBS). 2018.

LAVALLE, S.; LESSER, E.; SHOCKLEY, R.; HOPKINS, M.S.; KRUSCHWITZ, N. Big data, analytics and the path from insights to value. **MIT Sloan Management Review**, v. 52, n. 2, p. 21-32, 2011.

LEE, A. H. I.; WANG, W.-M.; LIN, T.-Y. An evaluation framework for technology transfer of new equipment in high technology industry. **Technological Forecasting and Social Change,** v. 77, n. 1, p. 135-150, Jan 2010.

LEE, I.; LEE, K. The internet of things (IoT): applications, investments, and challenges for enterprises. **Businnes Horizons,** v.58, n.4, p.431-440, 2015.

LEE, J.; BAGHERI, B.; KAO, H. A Cyber-physical Systems Architecture for Industry 4.0-based Manufacturing Systems. **Manufacturing Letters**, v.3, p.18-23, 2015.

LEE, J.; HAN, S.; YANG, J. Construction of a Computer-Simulated Mixed Reality Environment for Virtual Factory Layout Planning. **Computers in Industry,** v.62, p.86-98, 2011.

LEE, J.; LAIRA, E.; BAGHERI, B.; KAO, H. Recent advances and trends in predictive manufacturing systems in big data environment. **Manufacturing Letters,** v.1, n.1, p.38-41, 2013.

LEMA, A.; LEMA, R. Low-carbon innovation and technology transfer in latecomer countries: Insights from solar PV in the clean development mechanism. **Technological Forecasting and Social Change**, v. 104, p. 223-236, 2016.

LEMA, R.; HANLIN, R.; HANSEN, U. E.; NZILA, C. Renewable electrification and local capability formation: Linkages and interactive learning. **Energy Policy**, v. 117, p. 326-339, 2018.

LEMOS, D.; FERRAZ, S. University-industry interaction in Santa Catarina: evolutionary phases, forms of interaction, benefits, and barriers. **Journal of Administration and Innovation**, v. 14, n. 1, p. 16-29, 2017.

LEYDESDORFF, L. The triple helix, quadruple helix, and an N-tuple of helices: explanatory models for ana- lyzing the Knowledge-based economy? **Journal of the Knowledge Economy**, v.3, p.25-35, 2011.

LI, F.; ZHANG, S.; JIN, Y. Sustainability of University Technology Transfer: Mediating Effect of Inventor's Technology Service. **Sustainability**, v. 10, n. 6, p. 2085, 2018.

LIAO, Yongxin. Past, present and future of Industry 4.0-a systematic literature review and research agenda proposal. **International journal of production research**, v. 55, n. 12, p. 3609-3629, 2017.

LIBONI, L. B.; CEZARINO, L.; JABBOUR, O.; OLIVEIRA, C. J. C.; STEFANELLI, N. O. Smart industry and the pathways to HRM 4.0: implications for SCM. **Supply Chain Management: An International Journal**, 2019.

LIN, Bou-Wen; BERG, Daniel. Effects of cultural difference on technology transfer projects: an empirical study of Taiwanese manufacturing companies. **International Journal of Project Management**, v. 19, n. 5, p. 287-293, 2001.

LIN, T. C. Compliance, technology, and modern finance. **Brook. J. Corp. Fin. Com. L.**, v.11, p.159, 2016.

LIN, Tzu-Chieh; WANG, Kung Jeng; SHENG, Margaret L. To assess smart manufacturing readiness by maturity model: A case study on Taiwan enterprises. **International Journal of Computer Integrated Manufacturing**, v. 33, n. 1, p. 102-115, 2020.

LINK, A.; SIEGEL, D.; BOZEMAN, B. An empirical analysis of the propensity of academics to engage in informal university technology transfer. **Industrial and Corporate Change**, v. 16, n. 4, p. 641-655, 2007.

LIU, H.; ESSER, L. J.; WHITING, K. Realising Rio principles through sustainable energy solutions: Application of small hydropower (SHP) in China and other developing countries. **International Journal of Technology Management & Sustainable Development**, v. 12, n. 3, p. 281-300, 2013.

LIU, Hengwei; LIANG, Xi. Strategy for promoting low-carbon technology transfer to developing countries: The case of CCS. **Energy Policy**, v. 39, n. 6, p. 3106-3116, 2011.

LORNA, U.; HE, W. How the internet of things can help knowledge management: a case study from the automotive domain. **Journal of Knowledge Management**, v.21, n.1, p.57-70, 2017.

LOVETT, J. C.; HOFMAN, P. S.; MORSINK, K.; TORRES, A. B.; CLANCY, J. S.; KRABBENDAM, K. Review of the 2008 UNFCCC meeting in Poznań. **Energy Policy**, v. 37, n. 9, p. 3701-3705, 2009.

LUCKE, D.; CONSTANTINESCU, C.; WESTKA˙MPER, E. Smart Factory - A Step towards the Next Generation of Manufacturing. M. Mitsuishi, K. Ueda, and F. Kimura (Eds.), Manufacturing Systems and Technologies for the New Frontier**, Springer London, London**, p. 115-118, 2008.

LYBÆK, R., ANDERSEN, J. Enhancing the sustainable development contribution of future CDM projects in Asia. **Progress in Industrial Ecology - An International Journal**, v. 7, n. 1, p. 6-34, 2010.

MACGREGOR, S.P.; MARQUES-GOU, P.; AND SIMON-VILLAR, A. Gauging readiness for the quadruple helix: a study of 16 European organisations. **Journal of the Knowledge Economy**, v.1, p.173-190, 2010.

MAGRUK, A. Uncertainty in the sphere of the industry 4.0-potential areas to research. **Business, Management and Education**, v. 14, n. 2, p. 275-291, 2016.

MAHMOOD, T.; MUBARIK, M. S. Balancing innovation and exploitation in the fourth industrial revolution: Role of intellectual capital and technology absorptive capacity. **Technological Forecasting and Social Change**, n. *160*, 2020.

MAIA, R. F.; MASSOTE, A. A.; LIMA, F. Innovative laboratory model based on partnerships and active learning. **Frontiers in Education Conference (FIE). IEEE**, p.1-5, 2017.

MAKAREWICZ-MARCINKIEWICZ, A. Strategies Against Technological Exclusion. The Contribution of the Sustainable Development Concept to the Process of Economic Inclusion of Developing Countries. **Problemy Ekorozwoju - Problems of Sustainable Development**, v. 8, n. 2, p. 67-74, 2013.

MAKRINI, I. E. Working with Walt: How a Cobot Was Developed and Inserted on an Auto Assembly Line. **IEEE Robotics Autom**, v.25, n.2, p.51- 58, 2018.

MANYUCHI, A. E. Outward foreign direct investment from South Africa's energy sector and the transfer of environmentally sound technologies to Uganda's energy sector. **African Journal of Science, Technology, Innovation and Development**, v. 9, n. 3, p. 303-314, 2017.

MARQUES, M.; AGOSTINHO, C.; ZACHAREWICZ, G.; JARDIM-GONÇALVES, R. Decentralized decision support for intelligent manufacturing in Industry 4.0. **Journal of Ambient Intelligence and Smart Environments**, v.9, n.3, p.299-313, 2017.

MCADAM, R.; MILLER, K.; MCADAM, M.; TEAGUE, S. The development of university technology trans- fer stakeholder relationships at a regional level: lessons for the future. **Technovation**, v.32, p.57-67, 2012.

MEHTA, C.; SHANKAR, U.; BANDOPADHYAY, T. K. Low Carbon Technologies for Our Cities of Future: Examining Mechanisms for Successful Transfer and Diffusion. **India Quarterly**, v. 72, n. 4, p. 410-422, 2016.

MENDOZA, Ximena Patricia López; SANCHEZ, David Santos Mauricio. A systematic literature review on technology transfer from university to industry. **International Journal of Business and Systems Research**, v. 12, n. 2, p. 197-225, 2018.

MERZ, M.; MANGINI, M. Contractng and remote engineering consulting services.**Proceedings of the eLEGAL 2002 European Conference on Legal Aspects of ICT Application in Project-Based Business,** v.3, n.4, 2002.

METTLER, T. Maturity Assessment Models: A Design Science Research Approach. **International Journal of Society Systems Science**, v.3, n.1-2, p.81-98, 2011.

MICHNA, A.; KMIECIAK, R. Open-mindedness culture, knowledge-sharing, financial performance, and industry 4.0 in smes. **Sustainability**, v. *12*, n. 21, p. 1-17, 2020.

MICHNIEWICZ, J.; REINHART, G. Cyber-Physical Robotics - Automated Analysis, Programming and Configuration of Robot Cells Based on Cyber-Physical-Systems. **Procedia Technology,** v.15, p.566-575, 2014.

MILLER, Kristel; MCADAM, Rodney; MCADAM, Maura. A systematic literature review of university technology transfer from a quadruple helix perspective: toward a research agenda. **R&D Management**, v. 48, n. 1, p. 7-24, 2018.

MIRANDA, Francisco Javier; CHAMORRO, Antonio; RUBIO, Sergio. Re-thinking university spin-off: A critical literature review and a research agenda. **The Journal of Technology Transfer**, v. 43, n. 4, p. 1007-1038, 2018.

MOEUF, A.; PELLERIN, R.; LAMOURI, S.; TAMAYO-GIRALDO, S.; BARBARAY, R. The industrial management of SMEs in the era of Industry 4.0. **International Journal of Production Research**, v.56, n.3, p.1118-1136, 2018.

MOGHADDAM, M.; NOF, S. Collaborative Service-component Integration in Cloud Manufacturing. **International Journal of Production Research**, 2017.

MORELL, L.; TRUCCO, M. A proven model to re- engineer engineering education in partnership with industry. **World Engineering Education Forum, Buenos Aires, Argentina,** p.14-18, 2012.

MORGERA, E.; NTONA, M. Linking small-scale fisheries to international obligations on marine technology transfer. **Marine Policy**, v. 93, p. 295-306, 2018.

MOTYL, B.; BARONIO, G.; UBERTI, S.; SPERANZA, D.; FILIPPI, S. How will Change the Future Engineers' Skills in the Industry 4.0 Framework? A Questionnaire Survey. **Procedia Manufacturing**, v.11, p.1501-1509, 2017.

MOYNE, J.; ISKANDAR, J. Big data analytics for smart manufacturing: case studies in semiconductor manufacturing. **PROCESSES,** v.5, n.3, 2017.

MUELLER, E.; CHEN, X. L.; RIEDEL, R. Challenges and requirements for the application of industry 4.0: a special insight with the usage of cyber-physical system. **Chinese Journal of Mechanical Engineering**, v.30, n.5, p.1050-1057, 2017.

MULAMULA, G.; AMADI-ECHENDU, J. An examination of the potential links between ICT technology transfer and sustainable development**. International Journal of Technology Management & Sustainable Development**, v. 16, n. 2, p. 119-139, 2017.

NAKANO, D.; MUNIZ Jr, J.; BATISTA Jr, E. D. Engaging Environments: Tacit Knowledge Sharing on the Shop Floor. **Journal of Knowledge Management** , v.17, n.2, p.290-306, 2013.

NIKKHOU, Shima; TAGHIZADEH, K.; HAJIYAKHCHALI, S. Designing a portfolio management maturity model (Elena). **Procedia-Social and Behavioral Sciences**, v. 226, p. 318-325, 2016.

NESTOR F.; ALEJANDRO G.; DALENOGARE, L. S.; AYALA, F. Industry 4.0 technologies: Implementation patterns in manufacturing companies. **International Journal of Production Economics**, v. 210, p. 15-26, 2019.

NHAMO, G. REDD+ and the global climate policy negotiating regimes: Challenges and opportunities for Africa. **South African Journal of International Affairs**, v. 18, n. 3, p. 385-406, 2011.

NIDHRA, S., YANAMADALA, M., AFZAL, W., TORKAR, R. Knowledge transfer challenges and mitigation strategies in global software development-A systematic literature review and industrial validation. **International Journal of Information Management,** 2013, v. 33, n. 2, p. 333-355.

O'BRIEN, G.; O'KEEFE, P.; ROSE, J. Energy, poverty and governance. **International Journal of Environmental Studies**, v. 64, n. 5, p. 605-616, 2007.

OCKWELL, D. G. Key policy considerations for facilitating low carbon technology transfer to developing countries. **Energy Policy, v. 36**, Brighton, p. 4104-4115, 2008.

OESTERREICH, Thuy Duong; TEUTEBERG, Frank. Understanding the implications of digitisation and automation in the context of Industry 4.0: A triangulation approach and elements of a research agenda for the construction industry. **Computers in industry**, v. 83, p. 121-139, 2016.

OSABUTEY, E.L.C.; JIN, Z. Factors influencing technology and knowledge transfer: configurational recipes for Sub-Saharan Africa. **Journal of Business Research**, v. 69, n. 11, p. 5390-5395, 2016.

OZTEMEL, Ercan; GURSEV, Samet. Literature review of Industry 4.0 and related technologies. **Journal of Intelligent Manufacturing**, v. 31, n. 1, p. 127-182, 2020.

PAGANI, R.; KOVALESKI, J.; RESENDE, L. Methodi Ordinatio: a proposed methodology to select and rank relevant scientific papers encompassing the impact factor, number of citation, and year of publication. **Scientometrics**, v.105, n.3, p.2109-2135, 2015.

PALACIOS-MARQUÉS, Daniel; PERIS-ORTIZ, Marta; MERIGÓ, José M. The effect of knowledge transfer on firm performance. **Management Decision**, 2013.

PATIL, S. K.; KANT, R. A fuzzy AHP-TOPSIS framework for ranking the solutions of knowledge management adoption in supply chain to overcome its barriers. **Expert Systems with Applications,** 2014, v. 41, n. 2, p. 679-693.

PARK, S. H.; SHIN, W. S.; PARK, Y. H.; LEE, Y. Building a New Culture for Quality Management in the Era of the Fourth Industrial Revolution. **Total Quality Management & Business Excellence,** v.28, n.9, p.934-945, 2017.

PARK, S.-H.; LEE, Y.-G. Perspectives on Technology Transfer Strategies of Korean Companies in Point of Resource and Capability Based View. **Journal of technology management & innovation,** v. 6, p. 161-184, 2011.

PARNPHUMEESUP, P.; KERR, S. A. Stakeholder preferences towards the sustainable development of CDM projects: Lessons from biomass (rice husk) CDM project in Thailand. **Energy Policy**, v. 39, n. 6, p. 3591-3601, 2011.

PAULK, Mark. Capability maturity model for software. **Encyclopedia of Software Engineering**, 2002.

PEARCE, J.; ALBRITTON, S.; GRANT, G.; STEED, G.; ZELENIKA, I. A New Model for Enabling Innovation in Appropriate Technology for Sustainable Development. **Sustainability: Science, Practice, and Policy**. v. 8. p. 42-53, 2012.

PEI, F.Q.; TONG, Y.F.; HE, F.; LI, D.B. Research on design of the smart factory for forging enterprise in the Industry 4.0 environment. **Mechanika,** v.23, n.1, 2017.

PEREIRA, A. C.; ROMERO, Fernando. A review of the meanings and the implications of the Industry 4.0 concept. **Procedia Manufacturing**, v. 13, p. 1206-1214, 2017.

PÉREZ, M. P.; SÁNCHES, A. M. The development of university spin-offs: early dynamics of technology transfer and networking. **Technovation**, v. 23, p. 823-831, 2003.

PERSSON, J.G. Current trends in product development. **Procedia CIRP,** v.50, p.378-383, 2016.

PESSL, Ernst; SORKO, Sabrina Romina; MAYER, Barbara. Roadmap Industry 4.0-implementation guideline for enterprises. **International Journal of Science, Technology and Society**, v. 5, n. 6, p. 193-202, 2017.

PETER, J.; RYAN; RICHARD, B.; WATSON. Research challenges for the internet of things: what role can OR play? **Systems,** v.5, n.1, p.17-24, 2017.

PFEIFFER, S. Robots, Industry 4.0 and humans, or why assembly work is more than routine work. **Societies,** v.6, n.2, 2016.

PHUSAVAT, Kongkiti; KESS, Pekka. Roadmap for knowledge sharing and transfer: sustaining outsourcing relationships. **International journal of innovation and learning**, v. 5, n. 5, p. 496-511, 2008.

PINARD, M. I. Need for Effective Technology Transfer to Ensure Sustainability of Otta Seal. **Transportation Research Record: Journal of the Transportation Research Board**, v. 2349, n. 1, p. 129-135, 2013.

POPP, D. International Technology Transfer, Climate Change, and the Clean Development Mechanism. **Review of Environmental Economics and Policy**, v. 5, n. 1, p. 131-152, 2011.

POSADA J.; CARLOS T.; IÑ I.; DAVID O.; DIDIER S.; AMICIS R.; PINTO, E. B.; EISERT, P.; DÖLLNER, J.; VALLARINO, I.. Visual computing as a key enabling technology for industrie 4. 0 and industrial internet. **IEEE Computer Graphics and Applications,** v.35, n. 2, p.26-40, 2015.

POSSELT, G.; BÖHME, S.; AYMANS, S.; HERRMANN, C.; KAUFFELD, S. Intelligent learning management by means of multi-sensory feedback. **Procedia CIRP**, v.54, p.77-82, 2016.

PREUVENEERS, D.; ILIE-ZUDOR, E. The intelligent industry of the future: a survey on emerging trends, research challenges and opportunities in Industry 4.0. **Journal of Ambient Intelligence and Smart Environments**, v.9, n.3, 2017.

PUEYO, A. The role of technology transfer for the development of a local wind component industry in Chile. **Energy Policy, v. 39**, p. 4274-4283. apr. 2011.

QIN, Jian; LIU, Ying; GROSVENOR, Roger. A categorical framework of manufacturing for industry 4.0 and beyond. **Procedia Cirp**, v. 52, p. 173-178, 2016.

RADZIWON, A.; BILBERG, A.; BOGERS, M.; MADSEN, E. S. The smart factory: exploring adaptive and flexible manufacturing solutions. **Procedia engineering**, v. 69, p. 1184-1190, 2014.

REIF, R.; SHIRLEY, A.J.; LIVERIS, L. Report To The President Accelerating U.S. **Advanced Manufacturing,** 2014.

RAHMAN, Azmawani Abd; BENNETT, David. Advanced manufacturing technology adoption in developing countries. **Journal of Manufacturing Technology Management**, 2009.

RAHMAN, Azmawani Abd; BROOKES, Naomi J.; BENNETT, David J. The precursors and impacts of BSR on AMT acquisition and implementation. **IEEE Transactions on Engineering Management**, v. 56, n. 2, p. 285-297, 2009a.

RAHMAN, Azmawani; BENNETT, David; SOHAL, Amrik. Transaction attributes and buyer-supplier relationships in AMT acquisition and implementation: the case of Malaysia. **International Journal of Production Research**, v. 47, n. 9, p. 2257-2278, 2009b.

RAI, Varun; SCHULTZ, Kaye; FUNKHOUSER, Erik. International low carbon technology transfer: Do intellectual property regimes matter? **Global Environmental Change**, v. 24, p. 60-74, 2014.

RAUCH, E.; LINDER, C.; DALLASEGA, P. Anthropocentric perspective of production before and within Industry 4.0. **Computers & Industrial Engineering**, v.139, p.105-644, 2020.

RAVINDRANATH, N. H.; BALACHANDRA, P. Sustainable bioenergy for India: Technical, economic and policy analysis. **Energy**, v. 34, n. 8, p. 1003-1013, 2009.

RICCARDO PALMARINI; JOHN AHMET ERKOYUNCU; RAJKUMAR ROY. An Innovate Process to Select Augmented Reality (AR) Technology for Maintenance. **Procedia CIRP**, v.59, p. 23-28, 2017.

RIEGE, A. Three-dozen knowledge-sharing barriers managers must consider. **Journal of Knowledge Management,** 2005, v. 9, n. 3, p. 18-35.

RIDGWAY, K.; CLEGG, C. W.; WILLIAMS, D. J. The Factory of the Future. Future of Manufacturing Project: Evidence Paper 29. **London: Government Office for Science**, 2013.

ROMERO, D.; BERNUS, P.; NORAN, O.; STAHRE, J.; FAST-BERGLUND, Å. The operator 4.0: Human cyber-physical systems and adaptive automation towards human-auto- mation symbiosis work systems. **Advances in production management systems. Initiatives for a sustainable world. APMS 2016. IFIP Advances in Information and Communication Technology,** p. 1-11, 2016.

RÜßMANN, M.; LORENZ, M.; GERBERT, P.; WALDNER, M., JUSTUS, J.; ENGEL, P.; HARNISCH, M. Industry 4.0: the future of productivity and growth in manufacturing industries. **Boston Consulting Group**, 2015.

RYMASZEWSKA, A.; HELO, P.; GUNASEKARAN, A. IoT powered servitization of manu- facturing-an exploratory case study. **International Journal of Production Economics**, v.192, p. 92-105, 2017.

SÁBATO, J.A,; BOTANA, N. Science and technology in the future development of Latin America. ***El pimsamiento latinoamericano en la problemática ciencia-tecnología-desarrollo.*** Buenos Aires, Editorial Paidos, 1975.

SALKIN, C.; ONER, M.; USTUNDAG, A.; CEVIKCAN, E. A Conceptual Framework for Industry 4.0. **Industry 4.0: Managing the Digital Transformation, Springer Series in Advanced Manufacturing**, p.3-23, 2018.

SANDERS, Adam; ELANGESWARAN, Chola; WULFSBERG, Jens P. Industry 4.0 implies lean manufacturing: Research activities in industry 4.0 function as enablers for lean manufacturing. **Journal of Industrial Engineering and Management**, v. 9, n. 3, p. 811-833, 2016.

SANTANA, A.; AFONSO, P.; ZANIN, A.; WERNKE, R. Costing models for capacity optimization in Industry 4.0: Trade-off between used capacity and operational efficiency. **Procedia Manufacturing**, v.13, p.1183-1190, 2017.

SARKODIE, S. A.; STREZOV, V. Effect of foreign direct investments, economic development and energy consumption on greenhouse gas emissions in developing countries. **Science of The Total Environment**, v. 646, p. 862-871, 2019.

SASTOQUE, P.; LLORENTE, R.; TOLEDO, G.; DE LACALLE, L. N.; RAMEZANI, F. TRLs 5-7 Advanced Manufacturing Centres, Practical Model to Boost Technology Transfer in Manufacturing. **Sustainability,** v. 11, n. 18, 2019.

SATHAYE, J., BOUILLE, D., BISWAS, D., CRABBE, P., GENG, L., HALL, D., IMURA, H., JAFFE, A., MICHAELIS, L., PESZKO, G., VERBRUGGEN, A., WORRELL, E., YAMBA, F., TOLMASQUIM, M., JANZEN, H. Barriers, opportunities, and market potential of technologies and practices. **Climate Change**, v. 5, p. 345-398, 2001.

SCHMIDT, R.; MÖHRING, M.; HÄRTING, R. C.; REICHSTEIN, C.; NEUMAIER, P.; JOZINOVIĆ, P. Industry 4.0 - Potentials for Creating Smart Products: Empirical Research Results. **Business Information Systems**, p. 16-27, 2015.

SCHNEIDER, Paul. Managerial challenges of Industry 4.0: an empirically backed research agenda for a nascent field. **Review of Managerial Science**, v. 12, n. 3, p. 803-848, 2018.

SCHNEPP, O. United States-China Technology Transfer, **Prentice-Hall, Englewood Cliffs**, NJ, 1990.

SCHUH, G.; ANDERL, R.; GAUSEMEIER, J.; TEN HOMPEL, M.; WAHLSTER, W. Industrie 4.0 maturity index. Managing the digital transformation of companies. **Munich: Herbert Utz**, 2017.

SCHUH, G.; POTENTE, T.; WESCH-POTENTE, C.; HAUPTVOGEL, A. Sustainable Increase of Overhead Productivity due to Cyber-Physical Systems, 2013.

SCHUMACHER, A.; EROL, S.; SIHN, W. A maturity model for assessing Industry 4.0 readiness and maturity of manufacturing enterprises. **Procedia CIRP**, v.52, p.161-166, 2016.

SCHUMACHER, Andreas; EROL, Selim; SIHN, Wilfried. A maturity model for assessing Industry 4.0 readiness and maturity of manufacturing enterprises. **Procedia Cirp**, v. 52, n. 1, p. 161-166, 2016.

SCHUMACHER, Andreas; NEMETH, Tanja; SIHN, Wilfried. Roadmapping towards industrial digitization based on an Industry 4.0 maturity model for manufacturing enterprises. **Procedia Cirp**, v. 79, p. 409-414, 2019.

SCHUMANN, M.; LEYE, S.; POPOV, A. Virtual reality models and digital engineering solutions for technology transfer. **Applied Computer Systems**, v.17, n.1, p.27-33, 2015.

SCREMIN, L.; ARMELLINI, F.; BRUN, A.; SOLAR-PELLETIER, L.; BEAUDRY, C. Towards a Framework for Assessing the Maturity of Manufacturing Companies in Industry 4.0 Adoption. In Analyzing the Impacts of Industry 4.0 in Modern Business Environments. **IGI Global**, pp. 224-254, 2018.

SECUNDO, Giustina; DE BEER, Christle; PASSIANTE, Giuseppina. Measuring university technology transfer efficiency: a maturity level approach. **Measuring Business Excellence**, 2016.

SHARP, M.; AK, R.; HEDBERG Jr, T. A Survey of the Advancing use and Development of Machine Learning in Smart Manufacturing. **Journal of Manufacturing Systems,** p.1-10, 2018.

SHELBOURN, M.; HASSAN, T.; CARTER, C. Legal and contractual framework for the VO. **Virtual Organizations, Springer, Boston, MA**, p. 167-176, 2005.

SHI, Q.; LAI, X. Identifying the underpinning of green and low carbon technology innovation research: A literature review from 1994 to 2010. **Technological Forecasting and Social Change**, v. 80, n. 5, p. 839-864, 2013.

SHUJING, Q. The Analysis on Barriers of Low Carbon Technology Transfer. **Energy Procedia,** v. 14, p. 1398-1403, 2012.

SIEGEL, D.S.; WALDMAN, D.; LINK, A. Assessing the impact of organizational practices on the relative productivity of university technology transfer offices: an exploratory study. **Research Policy,** v.32, n.1, p.27-48, 2003.

SIEMIENIUCH, C. E.; SINCLAIR, M. A.; deC HENSHAW, M. J. Global Drivers, Sustainable Manufacturing and Systems Ergonomics. **Applied Ergonomics,** v.51, p.104-119, 2015.

SIERRA, Lizbeth Magdalena Puerta; VILLAZUL, Sergio Javier Jasso. University-industry linkage and technology transfer management. **Journal of Management Development**, 2018.

SINGH, M.; ABHISHEK, B. Transfer of environmentally sound technologies: identifying the hierarchy and interdependence of barriers. 2013.

SILVA JR, A. C.; ANDRADE, J. C. S.; LEAO, E. B. S.; WU, D. D. Sustainable Development and Cleaner Technology in Brazilian Energy CDM Projects: Consideration of Risks. **Human and Ecological Risk Assessment: An International Journal**, v. 19, n. 5, p. 1338-1358, 2013.

SINHA, M.; JOCHEM, R.; GEERS, D.; HEINZE, P. Maturity measurement of knowledge-intensive business processes. **The TQM Journal**, 2011.

SISINNI, E.; SAIFULLAH, A.; HAN, S.; JENNEHAG, U.; GIDLUND, M. Industrial internet of things: Challenges, opportunities, and directions. **IEEE Transactions on Industrial Informatics**, v.14, n.11, p.4724-4734, 2018.

SIVATHANU, B.; PILLAI, R. Smart HR 4.0-how Industry 4.0 is disrupting HR. **Human Resource Management International Digest**, v. 26, n. 4, p. 7-11, 2018.

SJÖDIN, David R. Smart Factory Implementation and Process Innovation: A Preliminary Maturity Model for Leveraging Digitalization in Manufacturing Moving to smart factories presents specific challenges that can be addressed through a structured approach focused on people, processes, and technologies. **Research-Technology Management**, v. 61, n. 5, p. 22-31, 2018.

SPATH, D.; GERLACH, S.; HÄMMERLE, M.; KRAUSE, T.; SCHLUND, S. Produktionsarbeit Der Zukunft-Industrie 4.0 (Production Work of the Future-Industry 4.0). **Study Fraunhofer-Institut Für Arbeitswirtschaft Und Organisation (IAO)**, p. 451-456, 2013.

STARK, J. Product Lifecycle Management: Paradigm for 21 Century Product Realization. Springer-Verlag, London, UK, 2004.

STOCK, T.; OBENAUS, M.; KUNZ, S.; KOHL, H. Industry 4.0 as enabler for a sustainable development: A qualitative assessment of its ecological and social potential. **Process Safety and Environmental Protection**, v.118, p.254-267, 2018.

STOREY, M. A. Theories, Methods and Tools in Program Comprehension: Past, Present and Future. In Program Comprehension, 2005. IWPC 2005. **Proceedings. 13th International Workshop,** p. 181-191, 2005.

STRANGE, R.; ZUCCHELLA, A. Industry 4.0 global value chains and international business. **Multinational Business Review**, v.25, n.3, 2017.

SUNG, Tae Kyung; GIBSON, David V. Knowledge and technology transfer: levels and key factors. **International Conference on Technology Policy and Innovation (ICTPI),** 2000.

SUNG, Tae Kyung. Industry 4.0: a Korea perspective. **Technological forecasting and social change**, v. 132, p. 40-45, 2018.

SYBERFELDT, A.; DANIELSSON, O.; HOLM, M.; WANG, L. Visual Assembling Guidance Using Augmented Reality. **Procedia Manufacturing**, v.1, p.98-109, 2015.

SYBERFELDT, A.; HOLM, M.; DANIELSSON, O.; WANG, L.; BREWSTER, R. L. Support Systems on the Industrial Shop-floor of the Future - Operator's Perspective on Augmented Reality. **Procedia CIRP,** v.44, p.108-113, 2016.

THAMES, L.; SCHAEFER, D. Industry 4.0: an overview of key benefits, technologies, and challenges. **Thames, L. and Schaefer, D. (Eds), Cybersecurity for Industry 4.0, Springer, Heidelberg, pp**. 1-33, 2017.

THOBEN, Klaus-Dieter; WIESNER, Stefan; WUEST, Thorsten. "Industrie 4.0" and smart manufacturing-a review of research issues and application examples. **International journal of automation technology**, v. 11, n. 1, p. 4-16, 2017.

THRAMBOULIDIS, K.; CHRISTOULAKIS, F. UML4IoT-A UML-based approach to exploit IoT in cyber-physical manufacturing systems. **Computers in Industry,** v.82, 2016.

TIRTO, Teofilo; OSSIK, Yuriy; OMELYANENKO, Vitaliy. ICT Support for Industry 4.0 Innovation Networks: Education and Technology Transfer Issues. In: **Design, Simulation, Manufacturing: The Innovation Exchange**. Springer, Cham, 2019. p. 359-369.

TJAHJONO, B.; ESPLUGUES, C.; ARES, E.; PELAEZ, G. What Does Industry 4.0 Mean to Supply Chain? **Procedia Manufacturing,** v.13, p.1175-1182, 2017.

TOFAIL, S. A. M.; KOUMOULOS, E. P.; BANDYOPADHYAY, A.; BOSE, S.; O'DONOGHUE, L.; CHARITIDIS, C. Additive manufacturing: scientific and technological challenges, market update and opportunities. **Materials Today,** v.21, n.1, p.22-37, 2018.

TORRES, A.; DUTRÉNIT, G.; SAMPEDRO, J.; BECERRA, N. What are the factors driving university- industry linkages in latecomer firms: evidence from Mexico.

Science and Public Policy, v. 38, n. 1, p. 31-42, 2011.

TORVANGER, A.; SHRIVASTAVA, M. K.; PANDEY, N.; TØRNBLAD, S. H. A two-track CDM: improved incentives for sustainable development and offset production. **Climate Policy**, v. 13, n. 4, p. 471-489, 2013.

TOURK, K.; MARSH, P. The new industrial revolution and industrial upgrading in China: achievements and challenges. **Economic and Political Studies**, v. 4, n.2, p.187-209, 2016.

TRAPPEY, A. J. C.; TRAPPEY, C. V.; HAREESH GOVINDARAJAN, U.; CHUANG, A. C.; SUN, J. J. A Review of Essential Standards and Patent Landscapes for the Internet of Things: A Key Enabler for Industry 4.0. **Advanced Engineering Informatics**, 2016.

TRENCHER, G.; YARIME, M.; MCCORMICK, K. B.; DOLL, C. N. H.; KRAINES, S. B. Beyond the third mission: Exploring the emerging university function of co-creation for sustainability. **Science and Public Policy**, v. 41, n. 2, p. 151-179, 2013.

TVARONAVIČIENĖ, M.; ČERNEVIČIŪTĖ, J. Technology transfer phenomenon and its impact on sustainable development. **Journal of Security and Sustainability Issues**. v. 5, p. 87-97, 2015.

URBAN, F.; SICILIANO, G.; SOUR, K.; LONN, P. D.; TAN-MULLINS, M.; MANG, G. South-South Technology Transfer of Low-Carbon Innovation: Large Chinese Hydropower Dams in Cambodia. **Sustainable Development**, v. 23, n. 4, p. 232-244, 2015.

VAC, C.S.; FITIU, A. Building sustainable development through technology transfer in a Romanian university. **Sustainability**, v. 9, n. 11, p. 2042, 2017.

VALMOHAMMADI, C. Examining the perception of Iranian organizations on internet of things solutions and applications. **Industrial and Commercial Training,** v.48, n.2, p.104-108, 2016.

VAN DER GAAST, W.; BEGG, K.; FLAMOS, A. Promoting sustainable energy technology transfers to developing countries through the CDM. **Applied Energy**, v. 86, n. 2, p. 230-236, 2009.

VAN HOOREBEEK, Mark. Frederick Cottrell in the 1912 Journal of Industrial and Engineering Chemistry: laying the foundations of university technology transfer. **Industrial and Engineering Chemistry Research**, v. 43, n. 4, p. 839-846, 2004.

VERDOLINI, E.; BOSETTI, V. Environmental Policy and the International Diffusion of Cleaner Energy Technologies. **Environmental and Resource Economics**, v. 66, n. 3, p. 497-536, 2017.

WAGIRE, A. A.; JOSHI, R.; RATHORE, A. P. S.; JAIN, R. Development of maturity model for assessing the implementation of Industry 4.0: learning from theory and practice. **Production Planning & Control**, p. 1-20, 2020.

WALLACE, E.; RIDDICK, F. Panel on Enabling Smart Manufacturing. **State College**, 2013.

WAN, J.F.; YI, M.L.; LI, D.; ZHANG, C.H.; WANG, S.Y.; ZHOU, K.L. Mobile services for customization manufacturing systems: an example of Industry 4.0. **IEEE Access,** v.4, 2016.

WANG L.; TO´RNGREN, M.; ONORI, M. Current status and advance- ment of cyber-physical systems in manufacturing. **Journal of Manufacturing Systems**, v.37, p. 517-527, 2015.

WANG, S. Y.; WAN, F.; LI, D.; ZHANG, C. H. Implementing smart factory of Industry 4.0: an outlook. **International Journal of Distributed Sensor Networks**. v.12, n.1, 2016.

WANG, Y.; YU, Y.; CHEN, M.; ZHANG, X.; WIEDMANN, H.; FENG, X. Simulating Industry: A Holistic Approach for Bridging the Gap between Engineering Education and Industry. Part I: A Conceptual Framework and Methodology. **International Journal Of Engineering Education**, v.31, n. 1, A, p.165-173, 2015.

WEIGAND J. PrauseM, Industry 4.0 and object-oriented development: incremental and architectural change. **Journal of technology management & innovation,** v.11, n.2, p.104-110, 2016.

WEISS, A.; HUBER, A. User Experience of a Smart Factory Robot: Assembly line workers demand daptative robots. **5th International Symposium on New Frontiers in Human-Robot Interaction**, 2016.

WEYER, S.; SCHMITT, M.; OHMER, M.; GORECKY, D. Towards Industry 4.0-Standardization as the crucial challenge for highly modular, multi-vendor production systems. **Ifac-Papersonline**, v. 48, n. 3, p. 579-584, 2015.

WILAMOWSKI, B. Welcome to the IEEE Transactions on Industrial Informatics, a New Journal of the Industrial Electronics Society. **IEEE Transactions on Industrial Informatics,** v.1, n.1, p.1-2, 2005.

WITTENBERG, C. Human-CPS interaction - requirements and human-machine interaction methods for the Industry 4.0. **IFAC-Papersonline,** n.49-19, p.420-425, 2016.

WOHLERS, T.; CAFFREY, T. Wohlers Report 2013. Additive Manufacturing and 3D Printing State of the Industry, Annual Worldwide Progress Report. **Fort Collins, Colorado, USA: Wohlers Associates, Inc,** 2013.

WOLF, M.; SERPANOS, D. Safety and security of cyber-physical and internet of-things systems. **Proceedings of the IEEE,** v.105, n.6, 2017.

WONG, K.S.; KIM, M.H. Privacy protection for data-driven smart manufacturing-turing system. **International Journal of Web Services Research,** v.4, n.3, 2017.

WORRELL, Ernst. Technology transfer of energy efficient technologies in industry: a review of trends and policy issues. **Energy Policy**, v. 29, n. 1, p. 29-43, 2001.

WU, W.; LEE, Y. Developing global managers' competencies using the fuzzy DEMATEL method. **Expert systems with applications**, v.32, n.2, p.499-507, 2007.

XU, J.; HUANG, E.; HSIEH, L.; LEE, L. H.; JIA, Q. S.; CHEN, C. H. Simulation optimization in the era of Industrial 4.0 and the Industrial Internet. **Journal of Simulation,** v.10, n.4, p.310-320, 2016.

XU, L. D.; HE, W.; LI, S. Internet of Things in Industries: A Survey. **IEEE Transactions on Industrial Informatics,** v.10, n.4, p.2233-2243, 2014.

XU, L. Enterprise Integration and Information Architecture. **New York: CRC Press**, 2015.

XU, L.; CAI, L.; ZHAO, S.; GE, B. Editorial: Inaugural Issue. **Journal of Industrial Integration and Management,** v.01, n.01, 2016.

XU, Li Da; XU, Eric L.; LI, Ling. Industry 4.0: state of the art and future trends. **International Journal of Production Research**, v. 56, n. 8, p. 2941-2962, 2018.

YAN, J.; XIN, S.; LIU, Q.; XU, W.; YANG, L.; FAN, L.; WANG, Q. Intelligent supply chain integration and management based on cloud of things. **International Journal of Distributed Sensor Networks**, v.10, n.3, p.624839, 2014.

YAN, Z.; ZHANG, P.; VASILAKOS, A. V. A survey on trust management for internet of things. **Journal of network and computer applications**, v.42, p.120-134, 2014.

YANG, C.; LAN, S.L.; SHEN, W.M.; HUANG, G.Q.; WANG, X.B.; LIN, T.Y. Towards product customization and personalization in IoT-enabled cloud manufacturing. **Cluster Computing**, v.20, n.2, 2017.

YOUNG, R. I. M.; GUNENDRAN, A. G.; CUTTING-DECELLE, A. F.; GRUNINGER, M. Manufacturing Knowledge Sharing in PLM: A Progression towards the Use of Heavy Weight Ontologies. **International Journal of Production Research,** v.45, n.7, p.1505-1519, 2007.

YU, Z.; OUYANG, J.; LI, S.; PENG, X. Formal modeling and control of cyber-physical manufacturing systems. **Advances in Mechanical Engineering,** v.9, n.10, 2017.

YUAN, Z.H.; QIN, W.Z.; ZHAO, J.S. Smart manufacturing for the oil refining and petrochemical industry. **Engineering 3,** n.2, 2017.

YUE, X.; CAI, H.; YAN, H.; ZOU, C.; ZHOU, K. Cloud-Assisted Industrial Cyber-Physical Systems: An Insight. **Microprocessors and Microsystems,** v.39, p.1262-1270, 2015.

YUN, J. J.; JEONG, E.; LEE, Y.; KIM, K. The effect of open innovation on technology value and technology transfer: A comparative analysis of the automotive, robotics, and aviation industries of Korea. **Sustainability**, v.10, n.7, p.2459, 2018.

Yun, J. J., Liu, Z. Micro- and macro-dynamics of open innovation with a Quadruple-Helix model. **Sustainability**, v.11, n.12, 2019.

ZANGIACOMI, A.; SACCO, M.; PESSOT, E.; DE ZAN, A.; BERTETTI, M. A Perspective for the Implementation of a Path Towards the Factory of the Future: The Italian Case. **2018 IEEE International Conference on Engineering, Technology and Innovation (ICE/ITMC)**, pp. 1-9, 2018.

ZAWADZKI, P.; ZYWICKI, K. Smart product design and production control for effective mass customization in the Industry 4.0 concept. **Management and Production Engineering Review,** v.7, n.3, 2016.

ZHENG, X.; MARTIN, P.; BROHMAN, K.; XU, L. Cloud Service Negotiation in Internet of Things Environment: A Mixed Approach. **IEEE Transactions on Industrial Informatics,** v.10, n.2, p.1506-1515, 2014.

ZHONG, Y. R.; CHEN XU; CHAO CHEN; GEORGE, Q. HUANG. Big Data Analytics for Physical Internet-based Intelligent Manufacturing Shop Floors. **International Journal of Production Research**, 2017.

APPENDIX

APPENDIX A - STAGES OF THE SYSTEMATIC LITERATURE REVIEW (SBR)

STAGES OF THE SYSTEMATIC LITERATURE REVIEW (RBS)

Stage 1 - Determine Research Intent: The use of this systematic review methodology is intended to aid in the analysis of the influence between Technology Transfer and Industry 4.0. Thus, this was the research intent of this topic contained within this methodology.

Stage 2 - Preliminary exploratory research with keywords: In the second stage, a preliminary research was conducted with keywords that suited the study in two main axes: "industry 4.0" and "technology transfer" in the *Scopus, Science Direct* and *Web of Science* databases. These three databases were selected due to returning an amount of articles with relevant impact factor.

Step 3 - Definition and combination of the keywords: As the aim of this work is to analyze the influence between TT and I4.0, it was realized in Step 2 that in order to have a consistent study base, it is necessary to divide this search among 8 groups, according to their main subjects and keywords. In addition, the terms "Industry 4.0" and "Technology Transfer", have several variants, which must all be present in the searches. Next, the keyword combinations in each of the groups are presented.

Group 1 aims to identify the main characteristics of Industry 4.0, the most important factors and everything that is being studied. In other words, to identify its state-of-the-art. If the search was made only with the variants of I4.0, the amount of articles would be very high (35818 articles), a factor that would make it impossible to read all of them. For this reason, the combination of the I4.0 variants together with the literature review variants was used. In this way, the articles found present the main focuses that the literature has been bringing on this theme. Figure 37 presents this combination of Group 1. In the same way, it is necessary to identify the state-of-the-art of Technology Transfer. As in I4.0, the same search strategy was used. This combination is presented in Figure 38.

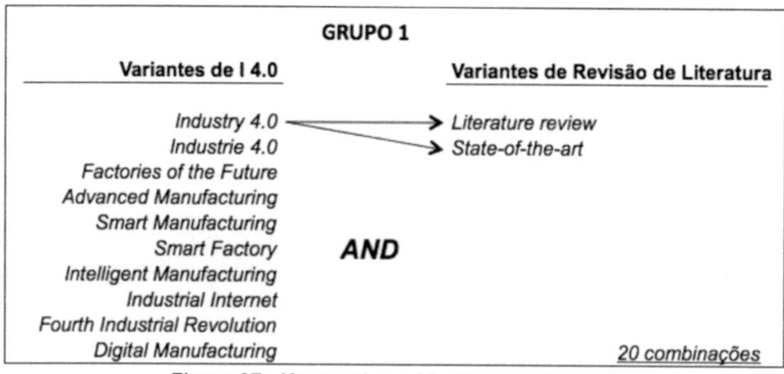

Figure 37 - Keyword combinations in Group 1
Source: Author (2021)

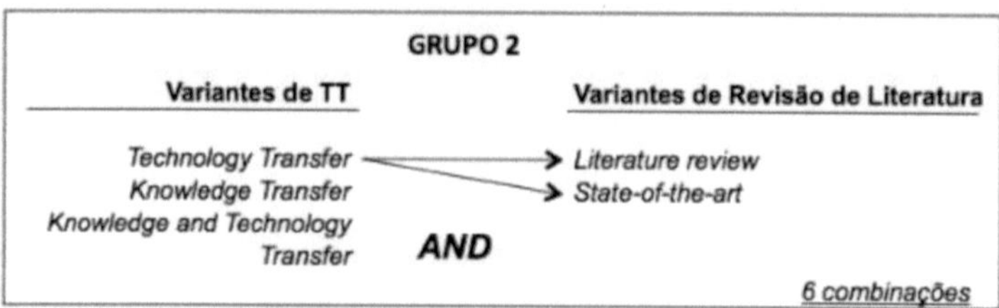

Figure 38 - Keyword combinations in Group 2
Source: Author (2021)

As it will be performed a maturity model in this study, it is necessary to have the knowledge of all existing maturity models in the literature of I4.0, and also those of TT. For this, Group 3 presents the combination of I4.0 variants together with the term "Maturity Model" (Figure 39), and Group 4 presents the combination of TT variants with "Maturity Model" (Figure 40).

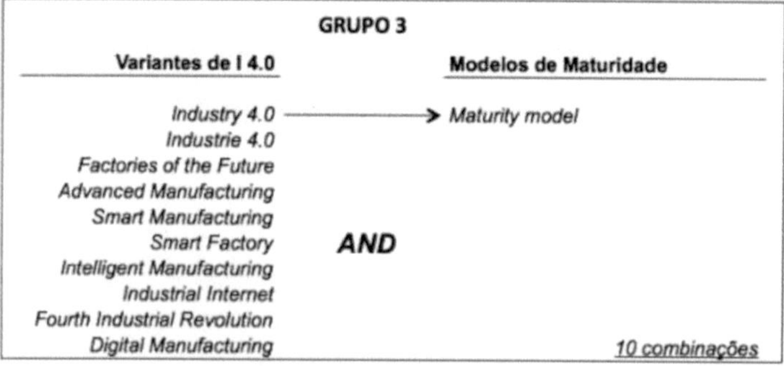

Figure 39 - Keyword combinations in Group 3
Source: Author (2021)

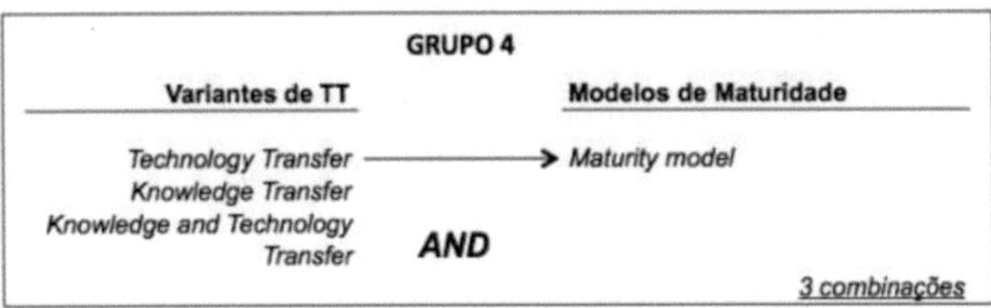

Figure 40 - Keyword combinations in Group 4
Source: Author (2021)

This work focuses primarily on the relationship between two factors: Industry 4.0 and Technology Transfer. Thus , Group 5 (Figure 41) presents the combination of all I4.0 and TT variants. Thus, it is possible to analyse all articles that study these two factors together.

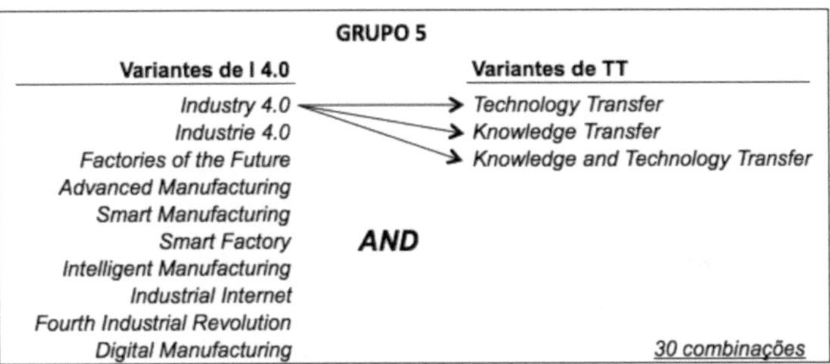

Figure 41 - Keyword combinations in Group 5
Source: Author (2021)

As the main objective of this work is to develop a maturity model of TT and I4.0, and thus analyse the influence of these two factors, it was necessary to search in the literature if there are already works with this focus. Group 6 presents the combination of the I4.0 variants, with the TT variants and the term "Maturity Model" (Figure 42). Group 7 presents the I4.0 variants, with the TT variants and the term "Influence" (Figure 43). Group 8 brings together all the keywords of this work: the variants of I4.0 and TT, along with the terms "Maturity Model" and "Influence" (Figure 44).

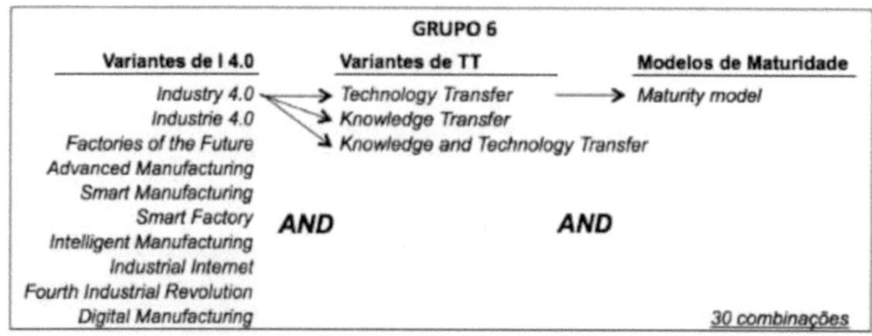

Figure 42 - Keyword combinations in Group 6
Source: Author (2021)

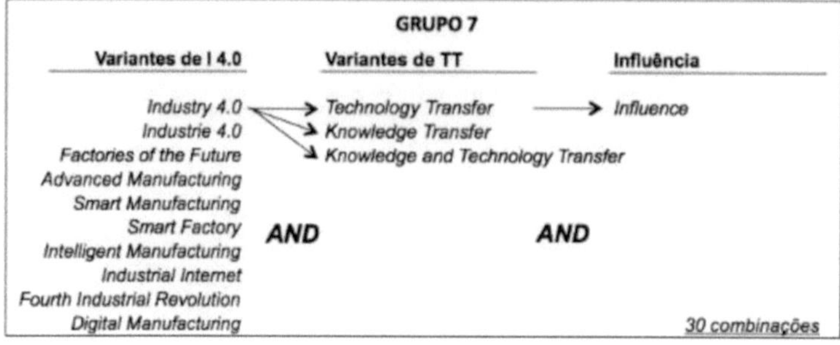

Figure 43 - Keyword combinations in Group 7
Source: Author (2021)

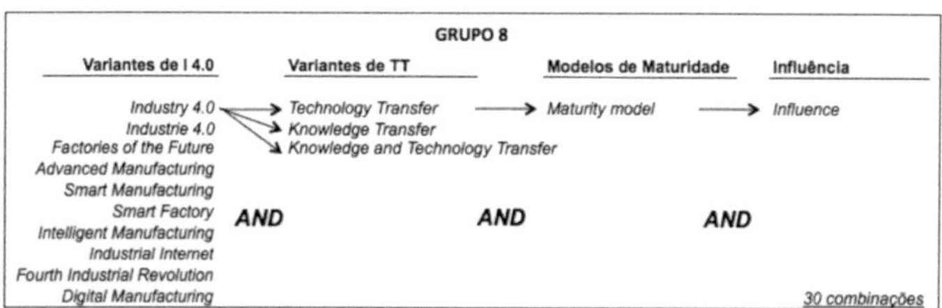

Figure 44 - Keyword combinations in Group 8
Source: Author (2021)

In all, 159 keyword combinations were performed in this search, in each of the three databases.

Step 4 - Final search in the databases: The final search was conducted using the keyword combinations described in the previous step in the *Web of Science, Science Direct* and *Scopus* databases to find a raw number of articles. A time cut-off of 21 years was used: jan/2000 < april/2021. Table 5 summarizes the initial results of this survey.

Table 5 - Number of articles found in each group

Groups	Keywords	Scopus	Science Direct	Web of Science	Total
Group 1	*("Industry 4.0" OR "Industrie 4.0" OR "Factories of the Future" OR "Advanced Manufacturing" OR "Smart Manufacturing" OR "Smart Factory" OR "Intelligent Manufacturing" OR "Industrial Internet" OR " Fourth Industrial Revolution" OR "Digital Manufacturing") AND ("Literature Review" OR "State-of-the-art")*	1032	593	565	**2190**
Group 2	*("Technology Transfer" OR "Knowledge Transfer" OR "Knowledge and Technology Transfer") AND ("Literature Review" OR "State-of-the-art")*	719	167	386	**1272**
Group 3	*("Industry 4.0" OR "Industrie 4.0" OR "Factories of the Future" OR "Advanced Manufacturing" OR "Smart Manufacturing" OR "Smart Factory" OR "Intelligent Manufacturing" OR "Industrial Internet" OR "Fourth Industrial Revolution" OR "Digital Manufacturing") AND ("Maturity Model")*	74	52	61	**187**
Group 4	*("Technology Transfer" OR "Knowledge Transfer" OR "Knowledge and Technology Transfer") AND ("Maturity Model")*	16	1	11	**28**
Group 5	*("Industry 4.0" OR "Industrie 4.0" OR "Factories of the Future" OR "Advanced Manufacturing" OR "Smart Manufacturing" OR "Smart Factory" OR "Intelligent Manufacturing" OR "Industrial Internet" OR " Fourth Industrial Revolution" OR "Digital Manufacturing") AND ("Technology Transfer" OR "Knowledge Transfer" OR "Knowledge and Technology Transfer")*	192	21	40	**253**
Group 6	*("Industry 4.0" OR "Industrie 4.0" OR "Factories of the Future" OR "Advanced Manufacturing" OR "Smart Manufacturing" OR "Smart Factory" OR "Intelligent Manufacturing" OR "Industrial Internet" OR " Fourth Industrial Revolution" OR "Digital Manufacturing") AND ("Technology Transfer" OR "Knowledge Transfer" OR "Knowledge and Technology Transfer") AND ("Maturity Model")*	0	0	1	**1**
Group 7	*("Industry 4.0" OR "Industrie 4.0" OR "Factories of the Future" OR "Advanced Manufacturing" OR "Smart Manufacturing" OR "Smart Factory" OR "Intelligent Manufacturing" OR "Industrial Internet" OR " Fourth Industrial Revolution" OR "Digital Manufacturing") AND ("Technology Transfer" OR "Knowledge Transfer" OR "Knowledge and Technology Transfer") AND ("Influence")*	1	1	1	**3**
Group 8	*("Industry 4.0" OR "Industrie 4.0" OR "Factories of the Future" OR "Advanced Manufacturing" OR "Smart Manufacturing" OR "Smart Factory" OR "Intelligent Manufacturing" OR "Industrial Internet" OR " Fourth Industrial Revolution" OR "Digital Manufacturing") AND ("Technology Transfer" OR "Knowledge Transfer" OR "Knowledge and Technology Transfer") AND ("Maturity Model") AND ("Influence")*	0	0	0	**0**

Source: Author (2021)

Therefore, a raw total of 3934 articles were surveyed to carry out the study in question. The information from these articles extracted in *BibTex* format was then taken to the reference management *software Mendeley Desktop©,* where the filtering process was initiated.

Step 5 - Filtering Processes: The raw number of articles was then filtered following the following criteria: "Exclusion of duplicates", "Exclusion of books and book chapters", "Exclusion by reading the titles" and "Exclusion by reading the abstract". Conference articles were not excluded due to the subject being recent and these types of articles may have relevant content for the research.

In the filtering by the "Duplicate Exclusion" criterion, through a tool available in the *Mendeley Desktop© software,* the raw number of articles was reduced to 2012 articles. These articles, now without duplicates, were filtered by the "Exclusion of books and book chapters" and "Exclusion by reading the titles" leading to the final number of the preliminary portfolio of 619 articles. The result of these filtrations can be better visualized in Table 6.

Table 5 - Number of articles in each group with the filtering steps

Groups	**No duplicates**	**Relevant articles** (exclusion of books and exclusion by reading titles and abstracts)
Group 1 - Industry 4.0	860	290
Group 2 - Technology Transfer	812	237
Group 3 - Maturity Models and Industry 4.0	98	34
Group 4 - Models of Maturity and Technology Transfer	21	7
Group 5 - Industry 4.0 and Technology Transfer	218	48
Group 6 - Industry 4.0, Technology Transfer and Maturity Models	1	1
Group 7 - Industry 4.0, Technology Transfer and Influence	2	0
Group 8 - Industry 4.0, Technology Transfer, Maturity and Influence Models	0	0

Source: Author (2021)

Thus, the 619 articles of the preliminary portfolio were extracted from the reference manager and washed into *Excel®* *software* so that they could be tabulated and analysed in the subsequent stage.

Stage 6 - Identification of the year of publication, impact factor and number of citations: In this stage, information was extracted from the 619 articles in the preliminary portfolio needed to carry out Stage 7. The year of publication was already contained in the files when extracted in *BibTex* format in Step 4. The impact factor was consulted through the relationship available on the *Clarivate Analytics website* and the number of citations of each article was acquired through *Google Scholar.*

Step 7 - Ordering of the articles according to *InOrdinatio*: This ordering step of the articles proposed in the methodology is carried out by calculating the *InOrdinatio* index through the following equation (1).

$$InOrdinato = (IF/1000) + \alpha * [10 - (Research\ Year - Publish\ Year)] + (\Sigma Ci)$$

(Equation 01)

Where, IF: Impact factor of the journal; α: Coefficient defined by the researcher; *Research Year*: Year in which the research was conducted; *Publish Year: Year* of publication of the article; Ci: Number of citations of the article in other works.

The value of α is attributed by the researcher and can vary between 1 and 10. The closer to 1, the less importance the researcher attributes to the year of publication of the articles, and the closer to 10, the greater the weight of the year of publication in the calculation. In this study, a value of 6 was attributed as this is an important criterion, but not of prime importance.

The articles in which the *InOrdinate* was negative were excluded from the portfolio. Subsequently, all these papers were analyzed through their titles and abstracts to see their relevance to the present research, so that these articles could finally be studied.

A total of 314 articles were selected to be analyzed and used as "raw material" in this research. Table 7 presents the number of articles pertaining to each group.

Table 6 - Number of articles in each group

Groups	Number of articles
Group 1 - Industry 4.0	107
Group 2 - Technology Transfer	139
Group 3 - Maturity Models and Industry 4.0	34
Group 4 - Models of Maturity and Technology Transfer	7
Group 5 - Industry 4.0 and Technology Transfer	27
Group 6 - Industry 4.0, Technology Transfer and Maturity Models	0
Group 7 - Industry 4.0, Technology Transfer and Influence	0
Group 8 - Industry 4.0, Technology Transfer, Maturity and Influence Models	0

Source: Author (2021)

Step 8 - Obtaining the articles in full: The 314 articles were acquired in full so that they could be analysed in Stage 9.

Stage 9 - Final reading and analysis of the articles: The final 314 articles were first studied in a bibliometric analysis. This analysis was performed based on the data found in the articles. Subsequently, a reading of these articles was carried out to extract information necessary for the composition of this work. Thus, a content analysis was performed to achieve the research objectives.

APPENDIX B - QUESTIONNAIRE 1

Questionnaire 1 - Level of Influence between I4.0 and TT dimensions

1) How much does the Human-Machine Interface influence Human Resources?
( ) 0 ( ) 1 ( ) 2 ( ) 3 ( ) 4

2) How much does the TT Process influence Human Resources?
( ) 0 ( ) 1 ( ) 2 ( ) 3 ( ) 4

3) How much does the I4.0 Process influence Human Resources?
( ) 0 ( ) 1 ( ) 2 ( ) 3 ( ) 4

4) How much does the Dissemination of Knowledge influence Human Resources?
( ) 0 ( ) 1 ( ) 2 ( ) 3 ( ) 4

5) How much do Technologies 4.0 influence Human Resources?
( ) 0 ( ) 1 ( ) 2 ( ) 3 ( ) 4

6) How much does the Organisational Culture influence the Human Resources?
( ) 0 ( ) 1 ( ) 2 ( ) 3 ( ) 4

7) How much do Business Models influence Human Resources?
( ) 0 ( ) 1 ( ) 2 ( ) 3 ( ) 4

8) How much do TT Modes influence Human Resources?
( ) 0 ( ) 1 ( ) 2 ( ) 3 ( ) 4

9) How much does Systems Integration influence Human Resources?
( ) 0 ( ) 1 ( ) 2 ( ) 3 ( ) 4

10) How much do Human Resources influence the Human-Machine Interface?
( ) 0 ( ) 1 ( ) 2 ( ) 3 ( ) 4

11) How much does the TT Process influence the Human-Machine Interface?
( ) 0 ( ) 1 ( ) 2 ( ) 3 ( ) 4

12) How much does the I4.0 process influence the Human-Machine Interface?
( ) 0 ( ) 1 ( ) 2 ( ) 3 ( ) 4

13) How much does the Dissemination of Knowledge influence the Human-Machine Interface?
( ) 0 ( ) 1 ( ) 2 ( ) 3 ( ) 4

14) How much do Technologies 4.0 influence the Human-Machine Interface?
( ) 0 ( ) 1 ( ) 2 ( ) 3 ( ) 4

15) How much does the Organisational Culture influence the Man-Machine Interface?
( ) 0 ( ) 1 ( ) 2 ( ) 3 ( ) 4

16) How much do Business Models influence the Human-Machine Interface?
( ) 0 ( ) 1 ( ) 2 ( ) 3 ( ) 4

17) How much do the TT Modes influence the Human-Machine Interface?
( ) 0 ( ) 1 ( ) 2 ( ) 3 ( ) 4

18) How much does Systems Integration influence the Human-Machine Interface?
( ) 0 ( ) 1 ( ) 2 ( ) 3 ( ) 4

19) How much do Human Resources influence the TT Process?
( ) 0 ( ) 1 ( ) 2 ( ) 3 ( ) 4

20) How much does the Human-Machine Interface influence the TT Process?
( ) 0 ( ) 1 ( ) 2 ( ) 3 ( ) 4

21) How much does the I4.0 Process influence the TT Process?
( ) 0 ( ) 1 ( ) 2 ( ) 3 ( ) 4

22) How much does Knowledge Dissemination influence the TT Process?
( ) 0 ( ) 1 ( ) 2 ( ) 3 ( ) 4

23) How much do Technologies 4.0 influence the TT Process?
( ) 0 ( ) 1 ( ) 2 ( ) 3 ( ) 4

24) How much does the Organisational Culture influence the TT Process?
( ) 0 ( ) 1 ( ) 2 ( ) 3 ( ) 4

25) How much do Business Models influence the TT Process?
( ) 0 ( ) 1 ( ) 2 ( ) 3 ( ) 4

26) How much do the TT Modes influence the TT Process?
( ) 0 ( ) 1 ( ) 2 ( ) 3 ( ) 4

27) How much does Systems Integration influence the TT Process?
( ) 0 ( ) 1 ( ) 2 ( ) 3 ( ) 4

28) How much do Human Resources influence the I4.0 Process?
( ) 0 ( ) 1 ( ) 2 ( ) 3 ( ) 4

29) How much does the Human-Machine Interface influence the I4.0 Process?
( ) 0 ( ) 1 ( ) 2 ( ) 3 ( ) 4

30) How much does the TT Process influence the I4.0 Process?
( ) 0 ( ) 1 ( ) 2 ( ) 3 ( ) 4

31) How much does the Dissemination of Knowledge influence the I4.0 Process?
( ) 0 ( ) 1 ( ) 2 ( ) 3 ( ) 4

32) How much do Technologies 4.0 influence the I4.0 Process?
( ) 0 ( ) 1 ( ) 2 ( ) 3 ( ) 4

33) How much does the Organizational Culture influence the I4.0 Process?
( ) 0 ( ) 1 ( ) 2 ( ) 3 ( ) 4

34) How much do Business Models influence the I4.0 Process?
( ) 0 ( ) 1 ( ) 2 ( ) 3 ( ) 4

35) How much do the TT Modes influence the I4.0 Process?
( ) 0 ( ) 1 ( ) 2 ( ) 3 ( ) 4

36) How much does Systems Integration influence the I4.0 Process?
( ) 0 ( ) 1 ( ) 2 ( ) 3 ( ) 4

37) How much do Human Resources influence the Dissemination of Knowledge?
( ) 0 ( ) 1 ( ) 2 ( ) 3 ( ) 4

38) How much does the Human-Machine Interface influence the Dissemination of Knowledge?
( ) 0 ( ) 1 ( ) 2 ( ) 3 ( ) 4

39) How much does the TT Process influence the Dissemination of Knowledge?
( ) 0 ( ) 1 ( ) 2 ( ) 3 ( ) 4

40) How much does the I4.0 Process influence the Dissemination of Knowledge?
( ) 0 ( ) 1 ( ) 2 ( ) 3 ( ) 4

41) How much do 4.0 Technologies influence the Dissemination of Knowledge?
( ) 0 ( ) 1 ( ) 2 ( ) 3 ( ) 4

42) How much does the Organizational Culture influence the Dissemination of Knowledge?
( ) 0 ( ) 1 ( ) 2 ( ) 3 ( ) 4

43) How much do Business Models influence Knowledge Dissemination?
( ) 0 ( ) 1 ( ) 2 ( ) 3 ( ) 4

44) How much do TT Modes influence Knowledge Dissemination?
( ) 0 ( ) 1 ( ) 2 ( ) 3 ( ) 4

45) How much does Systems Integration influence Knowledge Dissemination?
( ) 0 ( ) 1 ( ) 2 ( ) 3 ( ) 4

46) How much do Human Resources influence Technologies 4.0?
( ) 0 ( ) 1 ( ) 2 ( ) 3 ( ) 4

47) How much does the Human-Machine Interface influence Technologies 4.0?
( ) 0 ( ) 1 ( ) 2 ( ) 3 ( ) 4

48) How much does the TT Process influence Technologies 4.0?
( ) 0 ( ) 1 ( ) 2 ( ) 3 ( ) 4

49) How much does the I4.0 Process influence 4.0 Technologies?
( ) 0 ( ) 1 ( ) 2 ( ) 3 ( ) 4

50) How much does Knowledge Dissemination influence Technologies 4.0?
( ) 0 ( ) 1 ( ) 2 ( ) 3 ( ) 4

51) How much does the Organisational Culture influence the Technologies 4.0?
( ) 0 ( ) 1 ( ) 2 ( ) 3 ( ) 4

52) How much do Business Models influence Technologies 4.0?
( ) 0 ( ) 1 ( ) 2 ( ) 3 ( ) 4

53) How much do TT Modes influence Technologies 4.0?
( ) 0 ( ) 1 ( ) 2 ( ) 3 ( ) 4

54) How much does Systems Integration influence Technologies 4.0?
( ) 0 ( ) 1 ( ) 2 ( ) 3 ( ) 4

55) How much do Human Resources influence the Organisational Culture?
( ) 0 ( ) 1 ( ) 2 ( ) 3 ( ) 4

56) How much does the Human-Machine Interface influence the Organisational Culture?
( ) 0 ( ) 1 ( ) 2 ( ) 3 ( ) 4

57) How much does the TT Process influence the Organisational Culture?
( ) 0 ( ) 1 ( ) 2 ( ) 3 ( ) 4

58) How much does the I4.0 Process influence the Organisational Culture?
( ) 0 ( ) 1 ( ) 2 ( ) 3 ( ) 4

59) How much does the Dissemination of Knowledge influence the Organisational Culture?
( ) 0 ( ) 1 ( ) 2 ( ) 3 ( ) 4

60) How much do Technologies 4.0 influence the Organisational Culture?
( ) 0 ( ) 1 ( ) 2 ( ) 3 ( ) 4

61) How much do Business Models influence Organisational Culture?
( ) 0 ( ) 1 ( ) 2 ( ) 3 ( ) 4

62) How much do TT Modes influence Organisational Culture?
( ) 0 ( ) 1 ( ) 2 ( ) 3 ( ) 4

63) How much does Systems Integration influence the Organisational Culture?
( ) 0 ( ) 1 ( ) 2 ( ) 3 ( ) 4

64) How much do Human Resources influence Business Models?
( ) 0 ( ) 1 ( ) 2 ( ) 3 ( ) 4

65) How much does the Human-Machine Interface influence Business Models?
( ) 0 ( ) 1 ( ) 2 ( ) 3 ( ) 4

66) How much does the TT Process influence Business Models?
( ) 0 ( ) 1 ( ) 2 ( ) 3 ( ) 4

67) How much does the I4.0 Process influence Business Models?
( ) 0 ( ) 1 ( ) 2 ( ) 3 ( ) 4

68) How much does Knowledge Dissemination influence Business Models?
( ) 0 ( ) 1 ( ) 2 ( ) 3 ( ) 4

69) How much do Technologies 4.0 influence Business Models?
( ) 0 ( ) 1 ( ) 2 ( ) 3 ( ) 4

70) How much does the Organisational Culture influence the Business Models?
( ) 0 ( ) 1 ( ) 2 ( ) 3 ( ) 4

71) How much do TT Modes influence Business Models?
( ) 0 ( ) 1 ( ) 2 ( ) 3 ( ) 4

72) How much does Systems Integration influence Business Models?
( ) 0 ( ) 1 ( ) 2 ( ) 3 ( ) 4

73) How much do Human Resources influence TT Modes?
( ) 0 ( ) 1 ( ) 2 ( ) 3 ( ) 4

74) How much does Human-Machine Interface influence TT Modes?
( ) 0 ( ) 1 ( ) 2 ( ) 3 ( ) 4

75) How much does the TT Process influence the TT Modes?
( ) 0 ( ) 1 ( ) 2 ( ) 3 ( ) 4

76) How much does the I4.0 Process influence the TT Modes?
( ) 0 ( ) 1 ( ) 2 ( ) 3 ( ) 4

77) How much does the Dissemination of Knowledge influence the Modes of TT?
( ) 0 ( ) 1 ( ) 2 ( ) 3 ( ) 4

78) How much do Technologies 4.0 influence TT Modes?
( ) 0 ( ) 1 ( ) 2 ( ) 3 ( ) 4

79) How much does Organisational Culture influence TT Modes?
( ) 0 ( ) 1 ( ) 2 ( ) 3 ( ) 4

80) How much do Business Models influence TT Modes?
( ) 0 ( ) 1 ( ) 2 ( ) 3 ( ) 4

81) How much does Systems Integration influence TT Modes?
( ) 0 ( ) 1 ( ) 2 ( ) 3 ( ) 4

82) How much do Human Resources influence Systems Integration?
( ) 0 ( ) 1 ( ) 2 ( ) 3 ( ) 4

83) How much does the Human-Machine Interface influence Systems Integration?
( ) 0 ( ) 1 ( ) 2 ( ) 3 ( ) 4

84) How much does the TT Process influence Systems Integration?
( ) 0 ( ) 1 ( ) 2 ( ) 3 ( ) 4

85) How much does the I4.0 Process influence Systems Integration?
( ) 0 ( ) 1 ( ) 2 ( ) 3 ( ) 4

86) How much does Knowledge Dissemination influence Systems Integration?
( ) 0 ( ) 1 ( ) 2 ( ) 3 ( ) 4

87) How much do Technologies 4.0 influence Systems Integration?
( ) 0 ( ) 1 ( ) 2 ( ) 3 ( ) 4

88) How much does the Organizational Culture influence Systems Integration?
( ) 0 ( ) 1 ( ) 2 ( ) 3 ( ) 4

89) How much do Business Models influence Systems Integration?
( ) 0 ( ) 1 ( ) 2 ( ) 3 ( ) 4

90) How much do TT Modes influence Systems Integration?
( ) 0 ( ) 1 ( ) 2 ( ) 3 ( ) 4

APPENDIX C - Questionnaire 2

Questionnaire 2 - Maturity Level I4.0 and TT

If you are interested in receiving a diagnosis of your company's I4.0 maturity level, and actions that could be taken to reach a higher level, please leave your e-mail here:____

1. How are employees trained for their respective functions?
( ) No training occurs.
( ) The employees are trained only for their respective function.
( ) The higher hierarchical levels learn their function, and also the whole functioning of the organisation. The other employees are only trained for their respective function.
( ) All employees learn their function, and also the entire functioning of the organisation.
( ) The trainings are done with the help of technologies, such as augmented reality. All employees learn their function, and also the entire operation of the organisation.

2. Are there exchanges of professionals and/or researchers in the company?

( ) Does not occur.
( ) Very rarely.
( ) Occurs in some sectors.
( ) Occurs frequently in more than half of the sectors.
( ) Occurs very frequently, in all sectors.

3. Do employees have the technical and managerial skills required to implement Industry 4.0 actions?
( ) No.
( ) Very little. Only specific positions.
( ) They have on a medium scale, only some sectors.
( ) They have in more than half of the sectors.
( ) All employees have.

4. Is there interdisciplinarity in the working groups?
( ) No.
( ) Very rarely.
( ) It exists in some sectors.
( ) Exists in more than half of the sectors.
( ) It exists in all groups in all sectors.

5. Does the organisation carry out the necessary training for the acquisition of technical and managerial skills related to the concepts and technologies of Industry 4.0?
( ) Does not have.
( ) It is in the initial implementation phase.
( ) Partially in some sectors.
( ) Has in more than half of the sectors.

( ) All employees are being trained on I4.0.

6. Do the collaborators have autonomy and freedom for creation?

( ) No.
( ) Very little. Only top management.
( ) Partially in some sectors.
( ) Has in more than half of the sectors.
( ) All employees are being trained on I4.0.

7. Do employees master and have all the necessary knowledge in the technology they use?

( ) No.
( ) Master only the basics.
( ) A small scale of employees fully master the technologies they use.
( ) More than half of the employees are fully proficient in the technologies they use.
( ) All employees fully master the technologies they use.

8. Are employees open to new technologies?

( ) No.
( ) Very rarely.
( ) Only in some sectors.
( ) More than half of the employees are open to new technologies, regardless of their hierarchical level.
( ) All employees are open to new technologies.

9. Is there a specific R&D sector within the company?

( ) No.
( ) Only exists in units outside Brazil.
( ) Only exists in the Brazilian matrix.
( ) It is in the initial implementation phase in all units.
( ) Yes. All the company's units have an R&D sector.

10. Does the company acquire technologies and/or resources from third parties?

( ) No.
( ) Very rarely.
( ) Occurs on a small scale.
( ) Occurs frequently.
( ) Occurs very often.

11. Does the company have TT facilitators (such as consultants, technology transfer offices, development agencies...)?

( ) No.
( ) Very rarely.
( ) Possesses in some cases.
( ) Has many times.
( ) Always. Has contracts with TT facilitators.

12. Does the company transfer technologies to other companies / institutions?

( ) No.
( ) Very rarely.
( ) Occurs on a small scale.
( ) Occurs frequently.

( ) Occurs very often.

13. Does the company have flexibility in producing highly customised products?

( ) No.
( ) It is in the initial implementation phase.
( ) Partially in some specific products.
( ) Has more than half of the physical structure already implemented to meet customer needs.
( ) It has totally. The industry is able to meet the customised demand according to the client's request

14. What is the level of automation of the processes in the production/processing area of the company?

( ) From 0 to 20% of the processes are totally automated.
( ) 20 to 40% of the processes are totally automated.
( ) 40 to 60% of the processes are fully automated.
( ) 60 to 80% of the processes are totally automated.
( ) 80 to 100% of the processes are fully automated.

15. Does the company have an integrated system of sensors throughout the production process?

( ) Does not have.
( ) It is in the initial implementation phase.
( ) Partially in some sectors.
( ) Has in more than half of the sectors.
( ) Has totally in the industry.

16. Do industrial processes such as M2M carry out integrated information exchanges in real time?

( ) Do not carry out.
( ) It is in the initial implementation phase.
( ) Only in some sectors.
( ) In more than half of the sectors.
( ) They exchange information in real time throughout the company.

17. Are the production lines adaptable?

( ) No.
( ) Initial phase regarding the installation of the adaptable production line.
( ) Only in some production lines
( ) They are adaptable in more than half of the production lines.
( ) They are fully adaptable, being automatic and flexible for unexpected events.

18. Are there control systems that monitor the working conditions of the manufacturing machines?

( ) No.
( ) It is in the initial implementation phase.
( ) Only in some sectors / machines.
( ) Exists in more than half of the sectors / machines.
( ) It exists for all manufacturing machines in all sectors.

19. Does the Industry carry out intelligent maintenance?

( ) Does not carry out.
( ) Initial phase regarding the installation of Intelligent Maintenance resources.

( ) Performs in some sectors / production lines.
( ) Performs in more than half of the industrial processes.
( ) Performs totally in industrial processes.

20. Do the main processes have digital capabilities for modelling and simulating performance?
( ) No.
( ) It is in the initial implementation phase.
( ) Only in some processes.
( ) In more than half of the cases.
( ) Has in all processes.

21. Does the company have Patents and/or Intellectual Property?
( ) No.
( ) Has some, but has no interest in owning more.
( ) It has a reasonable quantity.
( ) Yes. However, only in some specific units of the company is it invested in order to obtain patents.
( ) Yes. The local company invests in researchers aiming at always having new patents.

22. Is the company interested and/or concerned with scientific publications?
( ) No.
( ) Very rarely.
( ) The company is interested, but does not invest in this area.
( ) The company is concerned and is initiating a process to increase scientific publications.
( ) Yes, the company invests so that more and more, new publications occur

23. Is knowledge shared among everyone in the company?
( ) No.
( ) Very rarely, only the necessary.
( ) The most basic and necessary knowledge that is shared on a small / medium scale.
( ) Knowledge is shared frequently, but it depends on the hierarchical level and/or sector.
( ) All knowledge is always shared, regardless of sector and/or hierarchical level.

24. Does the client have access to follow the stages of the production process?
( ) Does not have.
( ) It is in the initial implementation phase.
( ) Partially in some specific products.
( ) Has in more than half of the products.
( ) It has in all the company's products.

25. Are key business processes digitised and shared internally in the company and with other business partners?
( ) There is no digitalization of the processes.
( ) The processes are digitized, but remain in the sector. There is no sharing of this information.
( ) The processes are digitized and shared internally in the company. With the other business partners there is no sharing.
( ) The processes are digitised, and only in some cases are they shared internally in the company and with the other business partners.
( ) All processes are digitised and shared internally in the company and with the other business partners.

26. Does the company produce products with embedded and intelligent systems?
( ) No.
( ) Very rarely / Is in initial implementation phase.
( ) Produces small quantity.
( ) Produces frequently. More than half of the product portfolio are intelligent.
( ) All products have embedded and intelligent systems.

27. Is there traceability of the final product?
( ) Does not have.
( ) It is in the initial implementation phase.
( ) Partially in some specific products.
( ) Has in more than half of the products.
( ) It has in all the company's products.

28. Does the Industry use any Big Data resources?
( ) No.
( ) Initial phase regarding installation. The implementation was carried out in some specific processes in the Industry, generating only data agglomeration, without the resource of data analysis.
( ) It has partially. The configuration is still limited to the analysis of the large amount of data in some specific processes.
( ) It has, but it is still limited. The use of the Internet is not frequent, and the analysis of the large amount of data is not agile.
( ) Fully. The installation is by all processes where data capture is integrated between the sensors.

29. Does the Industry use Cloud Manufacturing?
( ) No.
( ) Initial use phase.
( ) It has partially. The resource is available in some specific processes in the Industry.
( ) Has in more than half of the company, with some limitations.
( ) Fully. All information is collected and stored in the cloud system.

30. Does the industry make use of any Information Security resources?
( ) No.
( ) Initial installation phase.
( ) It has partially. The resource is available in some specific processes.
( ) It has an advanced way. Some sectors already use and there are plans to expand.
( ) Fully. All information that is stored digitally is protected.

31. Does the industry work with augmented reality (AR - Augmented Reality) or virtual reality (VR - Virtual Reality)?
( ) No.
( ) Initial implementation phase.
( ) It has in specific sectors.
( ) Has in more than half of the sectors.
( ) It has in the whole company.

32. Does the industry use Additive Manufacturing (3D Printing) to compose the production line or for final products?
( ) No.

( ) Does not use, but is studying possibilities to use.
( ) Initial implementation phase.
( ) Uses in some specific lines.
( ) Uses in large quantity.

33. Does the industry have Autonomous Robots on the production line?
( ) No.
( ) Initial implementation phase.
( ) It has in specific sectors.
( ) Has in more than half of the sectors.
( ) It has in the whole company.

34. Is the equipment equipped with artificial intelligence technologies, which enable continuous improvement, enabling it to make decisions autonomously?
( ) No.
( ) Initial implementation phase.
( ) It has in specific sectors.
( ) Has in more than half of the sectors.
( ) It has in the whole company.

35. Are there consultancies in the company?
( ) No.
( ) Very rarely.
( ) Occurs with small frequency, only in some sectors.
( ) Occurs frequently.
( ) Occurs very often / always.

36. Does the company participate/make networking events?
( ) No.
( ) Very rarely.
( ) Participates with a regular frequency, but only managers and/or directors.
( ) Participates with a regular frequency, with varied hierarchical levels.
( ) Always participates, with varied hierarchical levels.

37. Does the company offer internships for students?
( ) Does not carry out.
( ) It is in the initial phase of implementation of internship programmes.
( ) Yes, but only voluntary internship.
( ) Yes, but only in some sectors.
( ) Yes. All sectors have at least one trainee.

38. Does the company have partnerships with master's and doctorate programmes?
( ) No.
( ) Does not perform yet, but is interested.
( ) Very rarely. Only when the masters / doctoral students come to us.
( ) Yes, we have partnerships with universities and participate in master's and doctoral projects.
( ) Yes. The company offers scholarships for masters and doctoral students to develop research of interest to the industry.

39. Do employees share their results in seminars/presentations/workshops?
( ) Never.
( ) Very rarely.

( ) Only internal presentations for your own sector.
( ) Only higher hierarchical levels present the results of their sector to the entire company.
( ) All employees share their results, presenting to the entire company.

40. Is the company open to change?

( ) There is no need to change anything in the company.
( ) Very little.
( ) Only some sectors and/or positions are open to changes. The others are very conservative.
( ) The board and top management positions are open to change. However, many employees are still afraid.
( ) Yes, all employees are prepared for changes.

41. Does the company frequently introduce innovations?

( ) Never.
( ) Very rarely.
( ) Presents on a small scale.
( ) Presents frequently, in more than half of the sectors.
( ) Always presents innovations, this is one of the company's goals.

42. Is there a central coordination for Industry 4.0 transformation actions?

( ) No.
( ) It is in the initial implementation phase.
( ) Only in the parent company.
( ) There is only one person responsible for I4.0 transformation actions in the local company.
( ) There is a team responsible for the implementation of Industry 4.0 in the local company.

43. Does the company have appropriate indicators to monitor the deployment of Industry 4.0 actions with realistic objectives and targets?

( ) No.
( ) It is in the initial implementation phase.
( ) It exists, but it is still very superficial.
( ) Exists, but only contemplates some sectors.
( ) It exists and is getting better and better.

44. Does the company plan and make the necessary investments to implement Industry 4.0 technologies?

( ) No.
( ) Very little.
( ) It is in the initial planning stage.
( ) There is planning, but not enough resources to implement 4.0 technologies at the moment.
( ) Yes. There is a planning and budget for the implementation of technologies 4.0.

45. Is there agile and digitised communication with customers for sales, services and information acquisition?

( ) No.
( ) It is in the initial implementation phase.
( ) It is carried out, but there are still restrictions and barriers that hinder this communication.
( ) It is carried out only with some customers and/or specific products.
( ) Yes, all communication with customers is digitalised.

46. Does the industry have marketing 4.0?

( ) Does not have.

( ) Initial implementation phase.
( ) Very little.
( ) It has partially.
( ) Fully owned.

47. Does the industry have partnerships with Universities?
( ) No partnership.
( ) The industry studies possibilities of partnerships.
( ) Initial phase of a partnership.
( ) Has a partnership with an institution.
( ) Has partnerships with several institutions.

48. What is the level of University-Company interaction?
( ) No interaction.
( ) Very low.
( ) Medium, there is a lot of bureaucracy and restrictions.
( ) High, there are still some restrictions.
( ) Very high. The university is fully open to industry, and vice versa.

49. Is there mutual trust in the buyer-supplier relationship?
( ) No.
( ) Very little.
( ) Reasonably.
( ) There is frequent trust.
( ) There is great confidence.

50. Is relevant information shared with other companies in the supply chain for quick decision-making?
( ) No.
( ) Very little.
( ) Reasonably.
( ) There is frequent sharing of information.
( ) All relevant information is shared.

51. Is there a transfer of knowledge between the different sectors of the company?
( ) No.
( ) Very little.
( ) Only in a superficial way.
( ) There is knowledge transfer between some sectors only.
( ) There is a great transfer of knowledge between all sectors of the company.

52. Is knowledge shared with suppliers?
( ) No.
( ) Very little.
( ) Reasonably.
( ) There is frequent sharing of knowledge.
( ) There is always knowledge sharing.

53. Is knowledge shared with customers?
( ) No.
( ) Very little.
( ) Reasonably.
( ) There is frequent sharing of knowledge.

( ) There is always knowledge sharing.

54. Is there technological knowledge transference from the R&D phase to the commercialisation of the product in the company?

( ) No.
( ) Very little.
( ) Only the knowledge that is of extreme importance is transferred.
( ) There is this vertical knowledge transfer, but not in a totalitarian way.
( ) All technological knowledge is shared from the R&D phase to the commercialisation of the product.

55. Is there technological knowledge transfer among the company's projects?

( ) No.
( ) Very rarely.
( ) Only in some sectors.
( ) There are in more than half of the sectors.
( ) All projects share technological knowledge among themselves.

56. Does the company exchange information in real time with its supply chain?

( ) No.
( ) It is in the initial implementation phase.
( ) There is exchange of information in real time, but there are still problems and failures in this communication.
( ) Only between some companies of the supply chain that there is exchange of information in real time.
( ) Yes, all information is exchanged in real time.

57. Between departments and hierarchical levels, is there an exchange of production information in real time?

( ) No.
( ) It is in the initial implementation phase.
( ) There is exchange of information in real time, but there are still problems and failures in this communication.
( ) Only between some sectors that there is exchange of information in real time.
( ) Yes, all information is exchanged in real time.

58. Does the Industry have any system, such as traceability, to manage the product life cycle (from raw material to disposal)?

( ) No.
( ) It is in the initial implementation phase.
( ) It has partially. Only some features of the product lifecycle management system are available.
( ) Has in more than half of the products.
( ) Fully equipped. It is possible to monitor in real time from the raw material to the withdrawal of the product from the market.

59. What is the company's level of connectivity?

( ) Very low.
( ) Low.
( ) Medium.
( ) High.

( ) Very high.

60. Is there bi-directional updating of information between the reis facilities and equipment with the digital copies?

( ) No.
( ) It is in the initial implementation phase.
( ) There are many flaws still, digital copies take time to be updated.
( ) Exists and works in more than half of the sectors / machines.
( ) It exists throughout the company, and the update is always in real time.

Printed by Books on Demand GmbH, Norderstedt / Germany